T0139856

Lecture Notes in Networks and Systems

Volume 963

The series "Lecture Notes in Networks and Systems" publishes the latest developments in Networks and Systems—quickly, informally and with high quality. Original research reported in proceedings and post-proceedings represents the core of LNNS.

Volumes published in LNNS embrace all aspects and subfields of, as well as new challenges in, Networks and Systems.

The series contains proceedings and edited volumes in systems and networks, spanning the areas of Cyber-Physical Systems, Autonomous Systems, Sensor Networks, Control Systems, Energy Systems, Automotive Systems, Biological Systems, Vehicular Networking and Connected Vehicles, Aerospace Systems, Automation, Manufacturing, Smart Grids, Nonlinear Systems, Power Systems, Robotics, Social Systems, Economic Systems and other. Of particular value to both the contributors and the readership are the short publication timeframe and the world-wide distribution and exposure which enable both a wide and rapid dissemination of research output.

The series covers the theory, applications, and perspectives on the state of the art and future developments relevant to systems and networks, decision making, control, complex processes and related areas, as embedded in the fields of interdisciplinary and applied sciences, engineering, computer science, physics, economics, social, and life sciences, as well as the paradigms and methodologies behind them.

Indexed by SCOPUS, INSPEC, WTI Frankfurt eG, zbMATH, SCImago.

All books published in the series are submitted for consideration in Web of Science.

For proposals from Asia please contact Aninda Bose (aninda.bose@springer.com).

Debasis Giri · Jaideep Vaidya · S. Ponnusamy ·
Zhiqiang Lin · Karuna Pande Joshi ·
V. Yegnanarayanan
Editors

Proceedings of the Tenth International Conference on Mathematics and Computing

ICMC 2024, Volume 2

 Springer

Editors
Debasis Giri
Department of Information Technology
Maulana Abul Kalam Azad University
of Technology
Kolkata, West Bengal, India

S. Ponnusamy
IIT Madras
Chennai, Tamil Nadu, India

Karuna Pande Joshi
The University of Maryland, Baltimore
County (UMBC)
Baltimore, USA

Jaideep Vaidya
Rutgers University
New Jersey, USA

Zhiqiang Lin
The Ohio State University
Columbus, USA

V. Yegnanarayanan
Kalasalingam Academy of Research
and Education
Srivilliputhur, Tamil Nadu, India

ISSN 2367-3370   ISSN 2367-3389 (electronic)
Lecture Notes in Networks and Systems
ISBN 978-981-97-2068-2   ISBN 978-981-97-2069-9 (eBook)
https://doi.org/10.1007/978-981-97-2069-9

This Springer imprint is published by the registered company Springer Nature Singapore Pte Ltd.
The registered company address is: 152 Beach Road, #21-01/04 Gateway East, Singapore 189721,
Singapore

If disposing of this product, please recycle the paper.

Committees

Chief Patron

Dr. K. Sridharan, Chancellor, Kalasalingam Academy of Research and Education, Tamil Nadu, India

Patrons

Smt. Dr. S. Arivalagi, Pro-Chancellor, Kalasalingam Academy of Research and Education, Tamil Nadu, India
Dr. S. Shasi Anand, Vice-President, (Academic), Kalasalingam Academy of Research and Education, Tamil Nadu, India
Er. S. Arjun Kalasalingam, Vice-President (Administration), Kalasalingam Academy of Research and Education, Tamil Nadu, India

Co-patrons

S. Narayanan, Vice Chancellor, Kalasalingam Academy of Research and Education, Tamil Nadu, India
V. Vasudevan, Registrar, Kalasalingam Academy of Research and Education, Tamil Nadu, India

General Chairs

P. K. Saxena, Scientific Consultant-Cyber Security, Government of India, Former Director, SAG, DRDO, New Delhi, India
P. D. Srivastava, IIT Bhilai, Raipur, India

Programme Chairs

Jaideep Vaidya, Rutgers University, USA
S. Ponnusamy, IIT Madras, Chennai, India
Zhiqiang Lin, The Ohio State University, USA
Karuna Pandey Joshi, The University of Maryland, Baltimore County (UMBC), USA
Debasis Giri, Maulana Abul Kalam Azad University of Technology, WB, India

Organizing Chair

Yegnanarayanan Venkataraman, Kalasalingam Academy of Research and Education, TN, India

International Advisory Committee

Alfred Menezes, University of Waterloo, Canada
Bhavani Thuraisingham, University of Texas, USA
Bimal Roy, ISI Kolkata, India
Chris Mitchell, Royal Holloway, University of London, UK
Dieter Gollmann, Hamburg University of Technology, Germany
Elisa Bertino, Purdue University, USA
Heinrich Begehr, Freie Universitat Berlin, Germany
Joel J. P. C. Rodrigues, University of Petroleum (East China), China
Kouichi Sakurai, Kyushu University, Japan
Lajos Hanzo, University of Southampton, Southampton, UK
Mahesh Kalyanaraman, Associate Vice-President, HCL, India
Mark Sepnaski, Baylor University, Waco, Texas, USA
Mohan Kankanhalli, National University of Singapore, Singapore
Mohammad S. Obaidat, University of Texas-Permian Basin, USA
Merouane Debbah, University of Texas-Permian Basin, USA
Moti Yung, Columbia University, USA
Oscar Castillo, Tijuana Institute of Technology, Mexico

Rajkumar Buyya, University of Melbourne, Australia
Rakesh M. Verma, University of Houston, USA
Sokratis Katsikas, Norwegian University of Science and Technology-NTNU, Norway
Valentina E. Balas, Aurel Vlaicu University of Arad, Romania
Veni Madhavan C. E., IISc Bangalore, India
Vishnu S. Pendyala, San Jose State University, USA

Members of the Organizing Committee

P. Deepalakshmi, Dean, SOC, KARE
C. Ramalingan, Dean, SAS, KARE
P. Sarasu, Director—International Relations, KARE
K. Karuppasamy, Professor of Mathematics, KARE
S. Balamurali, Senior Professor of Mathematics, KARE
S. Dhanasekaran, HOD, IT, KARE
K. Kartheeban, Head, CA and CSIT, KARE
M. Kameswari, HOD, Maths, KARE
K. Suthendran, Associate Professor (SG), IT, KARE
M. Jayalakshmi, Associate Professor, CSE, KARE
L. Sreenivasulu Reddy, Associate Professor, Maths, KARE
B. Reddappa, Associate Professor, Maths, KARE
T. Shanmughapriya, Assistant Professor, Maths, KARE
N. Sureshkumar, Professor, CSE, KARE
C. Balasubramanian, Associate Professor, CSE, KARE
N. C. Brintha, Associate Professor, CSE, KARE
AmuthaGuka, Associate Professor, CA, KARE
V. Baby Shalini, Associate Professor, IT, KARE
K. Maharajan, Associate Professor, CSE, KARE
Jane Rubel Angelina, Associate Professor, CSE, KARE
R. Murugeswari, Associate Professor, CSE, KARE
T. Sam PradeepRaj, Associate Professor, CSE, KARE
A. Parivazhagan, Associate Professor, CSE, KARE
V. Anusuya, Associate Professor, CSE, KARE
Abhishek Tripathi, Associate Professor, CSE, KARE

Members of the Design and Web Development

S. Suprakash and Prem Raja, School of Computing and his Team

Members of the Technical Programme Committee

Abhishek Kumar Singh, VIT University Chennai, India
Achyutha Krishnamoorthy, CMS College Kottayam, Kerala, India
A. Menaka Pushpa, VIT, Chennai, India
A. Vanav Kumar, National Institute of Technology, Arunachal Pradesh, India
Aditi Gangopadhyay, IIT Roorkee, India
Amit Prakash Singh, Guru Gobind Singh Indraprastha University, New Delhi, India
Amitabh Gyan Ranjan, MGCUB, Mahatma Gandhi Central University, Bihar, India
Amrit Pal, VIT, Chennai, India
Anita Pal, National Institute of Technology Durgapur, India
Anjana Gosain, Guru Gobind Singh Indraprastha University, Punjab, India
Anup Kumar Sharma, NIT Raipur, India
Anurag Goel, Delhi Technological University, New Delhi, India
Apu Kumar Saha, NIT Agartala, Agartala, Tripura, India
C. Arun Kumar, Amrita Vishwa Vidyapeetham, Coimbatore, India
Arvind K. R. Sinha, NIR, Raipur, India
Arya Kumar Bedabrata Chand, IIT Madras, India
Ashok Kumar Das, University of Calcutta, West Bengal, India
R. V. Aswiga, VIT, Chennai, India
Ajanta Das, Amity University Kolkata, India
Ali Ebrahimnejad, Qaemshahr Branch, Islamiz Azad University, Qaemshahr, Iran
Arif Ahmed Sk, University of Tromsø, Norway
Arvind Kumar Gupta, IIT Ropar, Punjab, India
Amitabh Gyan Ranjan, MGCUB, Mahatma Gandhi Central University, Bihar, India
Arshad Khan, Jamia Millia Islamia (A Central University) New Delhi, India
Abhijit Das, NIT Trichy, Thiruchirapalli, India
Ankita Vaish, Institute of Science Banaras Hindu University, India
Arvind Selwal, Central University of Jammu, India
Aquil Khan, Indian Institute of Technology Indore, India
A. Swaminathan, IIT Roorkee, India
Arvind Selwal, Central University of Punjab Bathinda, India
Atul, NIT, Trichy, India
Aditi Gangopadhyay, IIT Roorkee, India
A. K. Bedabrata Chand, IIT Madras, India
Amandeep Kaur, Central University of Jammu, India
Anshul Verma, Institute of Science, Banaras Hindu University, Varanasi, India
Abhijit Datta Banik, IIT Bhubaneswar, India
Ajit Das, Bodoland University Kokrajhar, Assam, India
Ajoy Kumar Khan, Mizoram University Aizawl, Mizoram, India
Aleksandr Poliakov, Sevastopol State University, India
Amit Banerjee, South Asian University, New Delhi, India
Amit Maji, IIT Kharagpur, India
Amiya Nayak, University of Ottawa, Canada

Ankit Rajpal, University of Delhi, Delhi, India
Anuj Jakhar, IIT Madras, India
Anupam Saikia, Indian Institute of Technology Guwahati, India
Arnab Patra, Indian Institute of Technology Kharagpur, India
Arun Kumar, National Institute of Technology, Rourkela, Odisha, India
A. Swaminathan, IIT Roorkee, India
Asish Bera, BITS Pilani, India
Atena Ghasemabadi, Esfarayen University of Technology, India
Ayaz Ahmad, NIT Patna, India
Ashok Kumar Das, International Institute of Information Technology, Hyderabad, India
Anoop Singh, IIT (BHU), Varanasi, India
Atasi Deb Ray, University of Calcutta, India
S. Ananda Kumar, VIT, Chennai, India
Annappa, NIT Karnataka, Surathkal, India
A. Senthil Thilak, NIT Karnataka, Surathkal, India
Alagu Manikandan Esakkimuthu, Cognizant, Hartford, USA
Adhvitha Premanand, NUS-National University of Singapore, Singapore
Anantha Narayanan, Accenture Technology Solutions, Dallas, USA
Arunkumar Pichaimuthu, Verizon, Concord, USA
Andavarayan Muthuraju, Syntel Ltd., Toronto, Canada
Anabayan, Cap Gemini, Atlanta, USA
Arun Kumar, Langara College, Vancouver, BC Richmond, Canada
Anandaraj Selvarangan, Tata Consultancy Services, Minneapolis, USA
Amutha, ABB Ltd., California City, USA
Akhil, Ericsson, San Francisco, USA
J. Abraham Stephen, TCS, Plano, USA
Arthy Paul Brito, Cardiff University, Wales, UK
Amit Kumar Verma, IIT Patna, India
Avijit Duary, Maulana Abul Kalam Azad University of Technology, West Bengal, India
Adriana Mihaela Coroiu, Babes-Bolyai University, Romania
Abd Raouf ChouikhaView Abd, Universite Paris-Sorbonne, Paris Nord, France
Angela Thompson, Governors State University, Greater Chicago, USA
Andrew Pownuk, The University of Texas at El Paso, USA
Abhimanyu Mukerji, Amazon, Bay Area, California, USA
Ashwin Viswanathan Kannan, Amazon Labs CA, USA
Balachander Palanisamy, Computer Sciences Corporation, USA
Balaji Krishnamoorthy, NTT Data UK, Bristol, UK
Bimal Mandal, IIT Jodhpur, India
Bipul Kumar Sarmah, Tezpur University, Napaam, Sonitpur, Assam, India
Bok-Min Goi, Universiti Tunku Abdul Rahman, Malaysia
Baskar Babujee, MIT Campus, Anna University, India
B. Sundaravadivoo, Alagappa University, Karaikudi, India
Bapan Ghosh, Indian Institute of Technology Indore, India

Dilip Kumar Maiti, Vidyasagar University, West Bengal, India
Dilpreet Kaur, Indian Institute of Technology Jodhpur, India
Djamal Foukrach, Hassiba Benbouali University of Chlef, Algeria
D. Bhargavi, NIT Warangal, India
Debashis Dutta, NIT Warangal, India
Debashis Nandi, National Institute of Technology, India
Debraj Giri, IIT, India
Deepak Ranjan Nayak, Malaviya National Institute of Technology Jaipur, Rajasthan, India
P. Deepalakshmi, Kalasalingam Academy of Research and Education, India
Devanayagam Palaniappan, Texas A&M University Corpus—Christi, USA
Dhananjoy Dey, Indian Institute of Information Technology Lucknow, India
Dinabandhu Pradhan, IIT (ISM) Dhanbad, India
Dipanwita Roy Chowdhury, IIT Kharagpur, India
Durga Charan Dalal, Indian Institute of Technology Guwahati, India
Debasish Bera, Indian Institute for Information Technology, Kalyani, India
Debasisha Mishra, NIT Raipur, India
Deepti Jain, Sri Venkateswara College, University of Delhi, India
C. Devadas Nayak, M.I.T., Manipal, India
Dhananjay Gopal, Guru Ghasidas Vishwavidyalaya (A Central University), Bilaspur (C.G.), India
Dharmendra Tripathi, National Institute of Technology, Uttarakhand, India
Edythe E. Weeks, Esq, Washington University, St. Louis, MO, USA
Floyd B. Hanson, University of Illinois, USA
Fateme Shahsavand, Malayer University, Malaysia
Fabienne Chaplais, Mathedu SAS, Paris, France
Fagen Li, University of Electronic Science and Technology of China, China
Fu-Hsing Wang, Chinese Culture University, China
Falguni Roy, NIT, Karnataka, Suratkal, India
A. Firos, Rajiv Gandhi University, Arunachal Pradesh, India
Gaurav Bhatnagar, Indian Institute of Technology Jodhpur, China
K. A. Germina, Central University of Kerala, Kerala, India
Gopal Chandra Shit, Jadavpur University, Kolkata, India
Gautham Singh, NIT, Trichy, India
Gururaj Mukarambi, Central University of Karnataka, India
Guru Prem Prasad, IIT Guwahati, India
G. S. Mahapatra, NIT, Puducherry, India
G. Malathi, VIT Chennai, India
Geetanjali Chattopadhyay, Malaviya National Institute of Technology Jaipur, India
T. Gireesh Kumar, Amrita Vishwa Vidyapeetham, Coimbatore, India
Gitanjali Chandwani, IIT Kharagpur, India
Gopal Chandra Shit, Jadavpur University, Kolkata, India
S. Gowrisankar, NIT, Patna, India
H. J. Gowtham, Manipal Academy of Higher Education, Manipal, India
Ganapathy Shanmugam Paramasivan, Medical University of Graz, Graz, Austria

Gaurav Kumar Singh, BE.services GmbH, Munich, Germany

Govindh Sankaran, Wolters Kluwer, Boston, USA

Ganesh Kumar, Judge Software, Charlotte, USA

Gerard Weiss, University of Maastricht, Germany

Halina Kwasnicka, Wroclaw University of Technology, USA

Hisao Ishibuchi, Osaka Prefecture University, Japan

H. Ramesh, Indian Institute of Technology Guwahati, Guwahati, India

Hari Vansh Rai Mittal, Indian Institute of Technology Palakkad, Kerala, India

H. Ramesh, Indian Institute of Technology Guwahati, Guwahati, India

Haowen Tan, Kyushu University, Japan

Hari Vansh RaiMittal, Indian Institute of Technology Palakkad, Kerala, India

Harshdeep Singh, DRDO, New Delhi, India

Heinrich Begehr, Free University Berlin, Germany, Germany

Hiranmoy Mondal, MAKAUT, India

Indranath Sengupta, IIT Gandhinagar, India

Ishapathik Das, IIT Tirupati, AP, India

Indivar Gupta, SAG, DRDO, India

Irani Hazarika, Gauhati University, Guwahati, India

Idrees Qasim, NIT, Srinagar, India

Ibrahim Venkat, Universiti Teknologi Brunei, Malaysia

Ingo Schiermeyer, TU Bergakademie Freiberg, Germany

Jos. M. Koomen, Memes Ltd., Utrecht, The Netherlands

Ji Wang, Ningbo University, China

Jey Veerasamy, The University of Texas at Dallas, USA

Joshva Rajkumar, Ministry for Primary Industries (MPI), Wellington, New Zealand

Joshua Gnanaruban, NCS Pte Ltd. (Singapore), Singapore

Jadav Das, MAKAUT, West Bengal, India

Jamuna Kanta Sing, Jadavpur University, West Bengal, India

Jaspal Singh Aujla, National Institute of Technology, Jalandhar, Punjab, India

JervinZen Lobo, St. Xavier's College Mapusa—Goa, India

Jothi Ramalingam MACS, NIT, Karnataka, Suratkal, India

Jyotsna Yadav, Guru Gobind Singh Indraprastha University, Punjab, India

Jagdish Prasad Maurya, Rajiv Gandhi University (A Central University), Arunachal Pradesh, India

Janakiraman, NIT, Trichy, India

Jaroslaw Adam Miszczak, Institute of Theoretical and Applied Informatics, Polish Academy of Sciences, Poland

Jaydeb Bhaumik, Jadavpur University, Kolkata, India

Jianting Ning, Singapore Management University, Singapore

Jugal K. Verma, IIT Bombay, India

J. Pavan Kumar, National Institute of Technology Tiruchirappalli, India

Jagdish Prakash, University of Johannesburg, South Africa

Jaideep Vaidya, Rutgers University, USA

Jayanta, Tezpur University, Assam, India

Jayanta Mukhopadhyay, Indian Institute of Technology Kharagpur, India

Jothi Ramalingam, MACS, NIT, Karnataka, Suratkal, India
Jugal Prajapat, Central University of Rajasthan, India
Jana Dittmann, Uni Magdeburg, India
Jyotismita Talukdar, Tezpur University, India
J. Christy Roja, St. Joseph's College, Trichy, India
K. Muthukumaran, VIT, Chennai, India
K. Somasundaram, Amrita Vishwa Vidyapeetham, Coimbatore, Tamil Nadu, India
Kartick Chandra Mondal, Jadavpur University, Kolkata, West Bengal, India
Karuna P. Joshi, University of Maryland, Baltimore County, USA
Kaushik Mondal, IIT, Ropar, Punjab, India
Kaushik Roy, West Bengal State University, Barasat, India
Kirankumar R. Hiremath, Indian Institute of Technology Jodhpur, India
Kunwer Singh Mathur, Dr. Hari Singh Gaur Viswavidyalaya Sagar, Central University of MP, India
Khurram Mustafa, JMI, New Delhi, India
K. Saraswathi, UCEK, JNTUK, India
Kadambari Raghuram, University College of Engineering Kakinada (Autonomous), Jawaharlal Nehru Technological University Kakinada, India
Kalpesh Kapoor, Indian Institute of Technology Guwahati, India
Kalyanbrat Medhi, Gauhati University, Assam, India
Khaleel Ahmed, Maulana Azad National Urdu University, Telangana, India
Khalid Mahmood, COMSATS University Islamabad, Sahiwal Campus, India
Kotaiah Bonthu, Central Tribal University of Andhra Pradesh, India
Kunwer Singh Mathur, Dr. Hari Singh Gaur Viswavidyalaya Sagar, Central University of MP, India
K. Palpandi, Malaviya National Institute of Technology (MNIT), Jaipur, Rajasthan, India
K. Saraswathi, UCEK, JNTUK, India
Kadambari Raghuram, University College of Engineering Kakinada (Autonomous), Jawaharlal Nehru Technological University Kakinada, India
Kamalika Bhattacharjee, NIT Trichy, India
Khalid Raza, Jamia Millia Islamia (Central University), New Delhi, India
Kolin Paul, IIT Delhi, India
Kushal Sharma, MNIT Jaipur, India
Kusum Sharma, National Institute of Technology, Uttarakhand, India
Karthikeyan Shenbagam, Link Systems, Houston, USA
Kamesh, Infosys, United States
Karthi Vicky, Hinduja Global Solutions, UK
Kouichi Sakurai, Kyushu University, Japan
Kumari Priyanka, University of Hohenheim, Germany
Kavitha Haldorai, Florida State University, USA
K. Somasundaram, Amrita Vishwa Vidyapeetham, Tamil Nadu, India
Longxiu Huang, Michigan State University, India
Lok Pati Tripathi, Indian Institute of Technology Goa Farmagudi, Ponda 403401, Goa, India

Lavanya Selvaganesh, Indian Institute of Technology (BHU), Varanasi, India
Lev Kazakovtsev, Siberian State Aerospace University, Russia
Maharage Nisansala Sevwandi Perera, ATR, Japan
Malay Banerjee, IIT Kanpur, India
Manikandan Rangaswamy, Central University of Kerala, Kerala, India
Manish Kumar Gupta, Guru Ghasidas Vishwavidyalaya, Bilaspur, India
Mario Larangeira, Tokyo Institute of Technology/IOHK, Japan
Md. Abu Talhamainuddin Ansary, IIT-Jodhpur, India
Meenakshi Thakur, Central University of Himachal Pradesh, India
Mohua Banerjee, IIT, Kanpur, India
Moumita Mandal, Indian Institute of Technology Jodhpur, Rajasthan, India
Mriganka Mandal, Indian Institute of Technology (IIT) Jodhpur, India
Muslim Malik, Indian Institute of Technology Mandi, India
Manoj Kumar Singh, Institute of Science, Banaras Hindu University Varanasi, India
Mritunjay Kumar Singh, IIT (ISM) Dhanbad, Dhanbad-826004, Jharkhand, India
Mohd. Arshad, Indian Institute of Technology Indore, Khandwa Road, Simrol, Indore, India
M. P. Pradhan, Sikkim University, Sikkim, India
M. Tiken Singh, Dibrugarh University Institute of Engineering and Technology, India
Madhumangal Pal, Vidyasagar University, India
Mahendra Kumar Gupta, IIT Bhubaneswar, India
Mahipal Jadeja, Malaviya National Institute of Technology, Jaipur, India
Manideepa Saha, NIT Meghalaya, India
Manmohan Vashisth, Indian Institute of Technology (IIT-JMU) Jammu, India
V. Mary Anita Rajam, Anna University, Chennai, India
Md. Obaidullah Sk, Aliah University, India
Megha Khandelwal, Central University of Karnataka, India
Muhammad Abulaish, South Asian University, New Delhi, India
T. Muthukumar, Indian Institute of Technology-Kanpur, India
M. Sethumadhavan, Amrita Vishwa Vidyapeetham, Coimbatore, India
Madhusudana Rao Nalluri, Amrita Vishwa Vidyapeetham, Coimbatore, India
Mahendra Kumar Gupta, IIT Bhubaneswar, India
Mahesh Shirole, VJTI-Mumbai, India
Malay Kule, IIEST, Shibpur, India
Manju Khari, Jawaharlal Nehru University, New Delhi, India
Md. Maqbul, National Institute of Technology Silchar, Assam, India
Mohammad Aslam Siddeeque, AMU, Aligarh, India
Mohammad Mueenul Hasnain, Kamala Nehru College University of Delhi, India
Munesh Meena, DSEU, New Delhi, India
M. Devakar, Visvesvaraya National Institute of Technology (VNIT) Nagpur, India
K. Manikandan, VIT, Vellore, India
Mahendra Pratap Singh, NIT Karnataka, Suratkal, India
Mohammed Hakim Jaffer Ali, SciLifeLab (Science for Life Laboratory), Stockholm, Sweden
Muthukrishnan Govindaraj, eGrove Systems, Sayreville, USA

Manivannan Karunanithi, Endera Systems LLC, McLean, USA
Mariappan Madasamy, HM Revenue and Customs, London, UK
Muthusamy Subash Rajasekaran, Tata Consultancy Services, Bloomington, USA
Mayan Sinha, Salesforce.com, USA
Muthu Rajathi, Tata Consultancy Services Ltd., Manchester, USA
Mohandoss Karuppiah, Cognizant Technology Solutions Pvt. Ltd., Virginia Beach, USA
Mihai Caragiu, Ohio Northern University, USA
Manimuthu Arunmozhi, Aston University, Birmingham, UK
Marcin Paprzycki, Systems Research Institute Polish Academy of Sciences, Warsaw, Poland
Nicholas Caporusso, Northern Kentucky University, USA
Navanietha Rathinam, American Society for Microbiology, Rapid City, USA
Nobin Saha, University of Bedfordshire, UK
Nazia Parveen, Aligarh Muslim University, Aligarh, India
V. Neelanarayanan, VIT, Chennai, India
Neha Kaushik, DSEU, New Delhi, India
Nemi Chandra Rathore, Central University of South Bihar, Gaya, India
Nirmal Kaur, Panjab University, Chandigarh, India
N. Balasubramani, NIT, Trichy, India
Neeraj Rathore, Indira Gandhi National Tribal University (IGNTU—A Central University), Amarkantak (M.P.), India
Neelesh S. Upadhye, IIT Madras, India
Neeraj Misra, Indian Institute of Technology, Kanpur, India
Niraj Khare, Carnegie Mellon University, USA
Nitu Kumari, IIT Mandi, India
N. R. Vemuri, University of Hyderabad, Hyderabad, India
N. Anbazhagan, Alagappa University, Karaikudi, India
Neetesh Saxena, Cardiff University, UK
Nemi Chandra Rathore, Central University of South Bihar, Gaya, India
Nesibe Yalçin, Erciyes University, elikgazi/Kayseri, Turkey
Om Prakash, IIT Patna, Bihar, India
C. Oswald, National Institute of Technology Tiruchirappalli, India
Om P. Suthar, Malaviya National Institute of Technology Jaipur, Rajasthan, India
Om Prakash Yadav, NIT, Hamirpur, India
Prakash Chelladurai, Max Planck Institute, Frankfurtam, Germany
Pruthvi Balachandra Kalyandurg, Swedish University of Agricultural Sciences, Uppsala, Sweden
Pon Janani Sugumaran, NUS-National University of Singapore, Singapore
Pratheep Kumar Reddy Yaddala, Target Corporation, Minneapolis, USA
Prakash Ramalingam, Civica, Singapore, Singapore
Pushpalatha Sekar, Cognizant Technology Solutions, San Jose, USA
S. Priya Dharshini, Syntel International Pvt. Ltd., United States
P. Muthu, NIT Warangal, India
Paras Ram, NIT, Kurukshetra, Haryana, India

Patil Shrishailappa Tatyasaheb, Vishwakarma Institute of Technology, Pune, India

V. Pattabiraman, VIT, Chennai, India

Piyali Debnath, NIT Agartala, Agartala, Tripura, India

Pradip Roul, VNIT, Nagpur, India

Pramod Kumar Goyal, Bhai Parmanand DSEU Shakarpur Campus-II, Delhi, India

Prashant Giridhar Shambharkar, Delhi Technological University, New Delhi, India

Prashant Kumar Srivastava, IIT Patna, Bihar, India

Prashant R. Nair, Amrita Vishwa Vidyapeetham, Coimbatore, India

Prem Prakash Mishra, NIT, Nagaland, India

Priti Kumar Roy, Jadavpur University, Kolkata, India

Priyanka Harjule, Malaviya National Institute of Technology (MNIT), Jaipur, Rajasthan, India

Projesh Nath Choudhury, IIT Gandhinagar, Gujarat, India

Purushottam Kar, IIT Kanpur, India

Pabitra Pal, Vidyasagar University, India

Prabhat Ranjan, Central University of South Bihar, Gaya, India

Prashant Kumar Srivastava, IIT Patna, Bihar, India

Pratibhamoy Das, IIT Patna, Bihar, India

Prodipto Das, Assam University (A Central University), Silchar, India

Prof. Dr. Jana Dittmann, University of Magdeburg, Germany

Projesh Nath Choudhury, IIT Gandhinagar, Gujarat, India

P. K. Parida, Central University of Jharkhand, Ranchi, India

P. Muthukumar, IIT Kanpur, India

P. P. Murthy, Guru Ghasidas Vishwavidyalaya (A Central University), Bilaspur (Chhattisgarh), India

Pablo Berna, CUNEF Universidad, Spain

Panchatcharam Mariappan, IIT Tirupati, India

Pankaj K. Das, Tezpur University, Sonitpur, Assam, India

Pawan Kumar, Indira Gandhi National Open University Maidan Garhi, New Delhi, India

Pradip Sasmal, IIT Jodhpur, India

Prasun Ghosal, Indian Institute of Engineering Science and Technology, Shibpur, India

Praveen Kumar Gupta, National Institute of Technology Silchar, India

Predrag Stanimirovic, University of Nis, Serbia

Punam Gupta, Dr. Hari Singh Gaur Viswavidyalaya Sagar, Central University of MP, India

Puneet Sharma, IIT Jodhpur, Rajasthan, India

Patitapaban Rath, Kalinga Institute of Industrial Technology (KIIT) Deemed to be University, Odisha, Bhubaneswar, India

Pramod Kewat, IIT, Dhanbad, India

P. K. Sahoo, Birla Institute of Technology and Science, Pilani Hyderabad Campus, India

P. Venkata Suresh, Indira Gandhi National Open University, New Delhi, India

Pushkar S. Joglekar, Viswakarma Institute of Technology, Pune, India

Promila Kumar, Gargi College (University of Delhi), India
Pedro Caceres, Fort Worth Metroplex, Dallas, USA
Radhakrishna Bhat, Manipal Institute of Technology, Manipal Academy of Higher Education, Manipal, Karnataka, India
S. Rajkumar, VIT, Chennai, India
Robert Richardson, Ravensbourne University London, UK
B. R. Rakshith, MIT, Manipal, India
Rajat Kumar Pal, University of Calcutta, India
G. Rajeshkumar, Bannari Amman Institute of Technology, TN, India
Raj Nandkeolyar, NIT, Jamshedpur, India
Romi Banerjee, IIT Jodhpur, India
Ranjit Kumar Upadhyay, Indian Institute of Technology (Indian School of Mines), Dhanbad, Jharkhand, India
R. Kalyanaraman, Annamalai University, Tamil Nadu, India
Rajat Kanti Nath, Tezpur University, Assam, India
Rajendra K. Ray, Indian Institute of Technology Mandi, India
D. Ranganatha, Central University of Karnataka, India
B. V. Rathish Kumar, IIT Kanpur, India
Rabinder Kumar Prasad, D.U.I.E.T. Dibrugarh University, Assam, India
Rafikul Alam, IIT Guwahati, Guwahati, India
Rahul Kumar Chawda, Assam University, Silchar, India
Rajendra Kumar Roul, Thapar Institute of Engineering and Technology, India
Rakesh Arora, IIT (BHU), Varanasi, India
Ranbir Sanasam, Indian Institute of Technology Guwahati, India
Ratikanta Behera, IISc Bangalore, India
Rohit Kumar Mishra, IIT Gandhinagar, Gujarat, India
Rupam Barman, Indian Institute of Technology Guwahati, Assam, India
Rifat Colak, Firat University, Türkiye
Reshma Rastogi, South Asian University, New Delhi, India
Ravi Subban, Pondicherry University, Puducherry, India
Ravi Kanth Asv, National Institute of Technology Kurukshetra, India
R. Eswari, NIT Trichy, India
R. Radha, University of Hyderabad, Hyderabad, India
R. Jagadeesh Kannan, Vellore Institute of Technology | Chennai Campus, Chennai, India
R. Kalyanaraman, Annamalai University, Tamil Nadu, India
R. Meher, S. V. National Institute of Technology, Surat, Gujarat, India
R. Suganya, VIT Chennai, India
Rabinder Kumar Prasad, D.U.I.E.T. Dibrugarh University, Assam, India
Rahul Kumar Chawda, Assam University, Silchar, India
Raj Kamal Maurya, SVNIT, Surat, Gujarat, India
Rajat Kanti Nath, Tezpur University, Assam, India
Rajat Tripathi, NIT, Jamshedpur, Jharkhand, India
Rajendra Kumar Roul, Thapar Institute of Engineering and Technology, India

Rajesh Ingle, International Institute of Information Technology, Naya Raipur, (IIIT NR), Chhattisgarh, India
Raksha Pandey, Guru Ghasidas Vishwavidyalaya, Bilaspur, India
Ramesh Kumar Vats, NIT Hamirpur, India
Ramesh Ragala, VIT Chennai, India
Ranjan Kumar Jana, Sardar Vallabhbhai National Institute of Technology (SVNIT) Surat, Gujarat, India
Ranjit Kumar Upadhyay, Indian Institute of Technology (Indian School of Mines), Dhanbad, Jharkhand, India
B. V. Rathish Kumar, IIT Kanpur, India
Reshma Rastogi (nee Khemchandani), South Asian University, New Delhi, India
Rifaqat Ali, NIT, Hamirpur, India
Ritu Agarwal, Malaviya National Institute of Technology Jaipur, Rajasthan, India
Rathina Kumar, University of Virginia, Charlottesville, USA
Raghunath Vel, University of Wolverhampton, UK
Rajaguru Paramasamy, H-E-B, Sr. PeopleSoft, San Antonio, USA
Ramprasad Renganathan, Omnitracs, North Atlanta, USA
Radhakrishnan Seenivasan, NTT DATA Americas, Pittsburgh, USA
Rajkumar Kathiresan, Cognizant Technology, Bentonville, USA
Revathi Balasubramanian, DFKI, Kaiserslautern, Germany
Raja Sekar, Cognizant Technology Solution, Toronto, Canada
Ram Bharadwaj, HCL Technologies, Ottawa, Canada
Rakesh Prabhakar, Cognizant, Winnipeg, Canada
Raja Saravanesh, Cognizant, Owings Mills, USA
Rajesh Kannan Karuppiah, Tata Consultancy Services, Jesus Martin, Mexico
Raja Lingam, Budapest University of Technology and Economics, Hungary
K. R. Renjith, Servian, Sydney, Australia
Ramsundar Kandasamy, Ericsson R&D, Germany
Raja Prabhu, Anya Consultancy Services, UK
B. Rushi Kumar, VIT Vellore, Tamil Nadu, India
Sri Padmavati. B, University of Hyderabad, Telangana, India
Sri Balaji Ponraj, The University of Sydney, Australia
Subathra Kannan, Biocon, Trichy, France
Swati Sharma, North Dakota State University, Fargo, USA
Satheesh Kumar, Nordic BioAnalysis AB, Stockholm, Sweden
Sannasi Nehru, Deloitte Consulting, Washington, USA
Sukumar Subburayan, CISCO, USA
Smitha Samuel, Sabre Inc. Irving, USA
Sangram Ray, National Institute of Technology Sikkim, India
Selvam Adaikkappan, Vistex, Chicago, USA
Saravanakumar Jagadeesan, Deloitte Consulting US, Austin, USA
Sujith Vijayakumar, Infosys Technologies Ltd., Dallas, USA
Sunil Ramkumar, Tata Consultancy Services, London, UK
Sadheesh Radhakrishnan, S-Cube Solutions Ltd. Location: London, UK
N. Shyam Sundar, Motorola Solutions, Plantation, USA

Siddharth Gopinath, Gainwell Technologies, Columbus, USA
Saravanan Sukumar, Cognizant Technology Solutions, Charlotte, USA
Saranya Kamarajan, TCS, USA
Sundar Ravanan, eHarmony, United States
Sowbhagya Lakshminarayanan, Greatness Packagers, Canada
Shanmuganathan, Tata Consultancy Services, United States
Suresh Raja, Tata Consultancy Services, United States
S. A. M. Rizvi, Jamia Millia Islamia, New Delhi, India
S. Amutha, Alagappa University, Karaikudi, India
S. Nithya Roopa, KIT, Coimbatore, India
S. R. Balasundaram, NIT, Trichy, India
Sakthi Prasad, National Institute of Technology, Arunachal Pradesh, India
Sam Johnson, NIT Karnataka, Suratkal, India
Sandeep Shinde, Vishwakarma Institute of Technology, Pune, India
Sanjeev Kumar, NSUT West Campus Jaffarpur Delhi, India
Saurabh Kumar Katiyar, NIT, Jalandar, Punjab, India
Shachi Sharma, South Asian University, New Delhi, India
Shafiqul Abidin, Aligarh Muslim University Aligarh, India
Shakir Ali, Aligarh Muslim University, Aligarh, India
Shanmugam Dhinakaran, Indian Institute of Technology Indore, India
Shraddha S. Suratkar, VJTI-Mumbai, India
Siddhartha Pratim Chakrabarty, Indian Institute of Technology Guwahati, Assam, India
Subrata Bera, National Institute of Technology Silchar, Assam, India
Subuhi Khan, Aligarh Muslim University, India
Sujoy Bhore, IIT Bombay, Mumbai, India
Sumit Kumar Debnath, NIT, Jamshedpur, Jharkhand, India
Sumit Nagpal, University of Delhi, India
Suraiya Jabin, Jamia Millia Islamia (Central University), New Delhi, India
Surendar Ontela, NIT, Mizoram, India
Susantha Maity, National Institute of Technology, Arunachal Pradesh, India
Sushil Kumar, S. V. National Institute of Technology Surat, Gujarat, India
T. R. Swapna, Amrita Vishwa Vidyapeetham, Coimbatore, India
S. Ponnusamy, IIT Madras, India
Saibal Pal, DRDO, New Delhi, India
Sangram Ray, National Institute of Technology Sikkim, India
Sanjay Mohanty, VIT Vellore, India
Santanu Sarkar, IIT Madras, India
Sarita Ojha, Indian Institute of Engineering Science and Technology Shibpur, West Bengal, India
Satrajit Ghosh, Aarhus University, Denmark
Sedat Akleylek, Ondokuz Mayis University, Turkey
P. Shaini, Central University of Kerala, Kerala, India
Sharanjeet Dhawan, NIIT University, Rajasthan, India
Sharmistha Adhikari, NIT Sikkim, India

Sk Hafizul Islam, IIIT Kalyani, India
Sokratis Katsikas, Norwegian University of Science and Technology, Norway
Sourav Mandal, XIM University Bhubaneswar, Odisha, India
Subhas Barman, Jalpaiguri Government Engineering College, West Bengal, India
Subhasis Dasgupta, University of California, San Diego, USA
Suprio Bhar, IIT Kanpur, India
Syed Abbas, IIT Mandi, India
Sandeep Singh Rawat, IGNOU, New Delhi, India
A. Sathishkumar, IIT Madras, India
Sanyasiraju, IIT Madras, India
S. P. Tiwari, Indian Institute of Technology (Indian School of Mines), Dhanbad, Jharkhand, India
Srinivas Kumar Vasana, Indian Institute of Technology Delhi, India
Srinivasa Rao Pentyala, Indian Institute of Technology (ISM) Dhanbad, India
Suchandan Kayal, National Institute of Technology Rourkela, Odisha, India
Sanjeev Singh, Indian Institute of Technology, Indore, MP, India
K. C. Srikantaiah, SJB Institute of Technology, Bengaluru, Karnataka, India
Sangita Jha, NIT Rourkela, India
Sairam Kaliraj, IIT Ropar, Punjab, India
Sreenivasulu Ballem, Central University of Karnataka Kalaburagi, India
Sanjay Kumar, Central University of Jammu, India
S. K. Pandey, IIT (BHU, Varanasi), India
S. P. Tiwari, Indian Institute of Technology (Indian School of Mines), Dhanbad, Jharkhand, India
S. K. V. Jayakumar, Pondicherry University, Pudhucherry, India
Sabyasachi Dutta, University of Calgary, Canada
Sabyasachi Pani, IIT Bhubaneswar, India
Sahana Prasad, BITS Pilani, Rajasthan, India
Saifur Rahman, Rajiv Gandhi University Doimukh, India
Saiyed Umer, ISI Kolkata, India
Saminathan Ponnusamy, IIT Madras, India
Santanu Saha Ray, National Institute of Technology Rourkela, Odisha, India
Sartaj Ul Hasan, Indian Institute of Technology Jammu, India
Sasmita Barik, IIT Bhubaneswar, India
Sujit Das, NIT Warangal, India
A. Sathish Kumar, IIT Madras, India
Satyanarayana Engu, NIT, Warangal, Andhra Pradesh, India
Sedat Akleylek, Ondokuz Mayis University, Turkey
Shailesh Kumar Tiwari, Indian Institute of Technology Patna, Bihar, India
Shibesh Kumar Jas Pacif, VIT, Vellore, India
Shripad M. Garge, Indian Institute of Technology, Mumbai, India
Shyamalendu Kandar, Indian Institute of Engineering Science and Technology, Shibpur, India
Siddhartha Pratim Chakrabarty, Indian Institute of Technology Guwahati, Assam, India

Siuli Mukhopadhyay, Indian Institute of Technology Bombay, Mumbai, India

Sivaram Ambikasaran, IIT Madras, India

Somesh Kumar, Indian Institute of Technology Kharagpur, India

Somnath Dey, Indian Institute of Technology Indore, India

Subinoy Chakraborty, Jadavpur University, Kolkata, West Bengal, India

Srinivasa Rao Kola, NIT, Karnataka, Suratkal, India

Subit Kumar Jain, NIT, Hamirpur, India

Srinivasu Bodapati, IIT Mandi, India

Subir Das, Indian Institute of Technology (BHU), Varanasi, India

Sudesh Rani, Punjab Engineering College, Deemed to be University, Chandigarh, India

Sudipta Majumder, Dibrugarh University Institute of Engineering and Technology (DUIET), Assam, India

Sujata Pal, IIT Ropar, India

K. Sumesh, IIT Madras, Chennai, India

S. Bose, College of Engineering, Anna University, India

Sujoy Bhore, IIT Bombay, India

Satya Bagchi, NIT Durgapur, India

Swaleha Zubair, Aligarh Muslim University, India

S. Gandhiya Vendhan, Bharathiar University, Coimbatore, India

Sujit Das, NIT, Warangal, India

Shafik, Nanjing University of Information Science and Technology, China

Scott Baldridge, Baton Rouge, LA, USA

Tarak Gaber, The University of Salford, UK

Terry Kaufman, Institute for Mathematics and Computer Science, Fort Lauderdale, FL, USA

Triloki Nath, Dr. Hari Singh Gour Vishwavidyalaya, Sagar, MP, India

Taqseer Khan, Jamia Millia Islamia, Jamia Nagar, New Delhi, India

Tanmoy Maitra, KIIT University, Bhubaneswar, India

Tapas Chatterjee, IIT Ropar, Punjab, India

Tarun Yadav, Defence Research and Development Organisation, New Delhi, India

Tingwen Huang, Texas A&M University, USA

Tuhina Mukherjee, Indian Institute of Technology Jodhpur, India

T. Subbulakshmi, VIT, Chennai, India

Tamal Pramanick, NIT Calicut, Kozhikode, Kerala, India

Tanweer Jalal, NIT, Srinagar, India

Tarni Mnadal, NIT, Jamshedpur, Jharkhand, India

Thomas George, Missing Link Technologies Ltd., Moncton, Canada

Thirumaran Pathakkam Mannai, Elavon, Inc. Knoxville, USA

Ujwal Warbhe, NIT, Srinagar, India

Usha Rani, Sri Venkateswara University, Tirupati, Andhra Pradesh, India

Uaday Singh, Indian Institute of Technology Roorkee, Roorkee-247667 (Uttarakhand), India

Ushnish Sarkar, Netaji Subhas Open University, India

Utpal Roy, Siksha-Bhavana Visva-Bharati Santiniketan, Birbhum, WB, India

Udayan Prajapati, St. Xavier's College, Navrangpura, Ahmedabad, India
Vilem Novak, University of Ostrava, Czech Republic
Vinod Kumar, PGDAV College, University of Delhi, Nehru Nagar, New Delhi, India
V. Balakumar, National Institute of Technology Puducherry, Karaikal, India
Vishnu Narayan Mishra, Indira Gandhi National Tribal University, Madhya Pradesh, India
V. V. Subrahmanyam, IGNOU, New Delhi, India
V. Shanthi, NIT Trichy, Tamil Nadu, India
V. D. Ambeth Kumar, Mizoram University, Mizoram, India
Vipindev Adat Vasudevan, Massachusetts Institute of Technology, Cambridge, USA
V. Shanthi, NIT Trichy, Tamil Nadu, India
Vaibhav Dhore, VJTI-Mumbai, India
R. Vedhapriyavadhana, VIT, Chennai, India
Vijayakumar Ramakrishnan, NIT, Calicut, Kerala, India
K. V. Vijayashree, Anna University, Chennai, India
Vinay Singh, NIT Mizoram, India
Vipindev Adat Vasudevan, Massachusetts Institute of Technology, Cambridge, India
Venkatesh Babu Nattamai Balakrishnan, Diconium Digital Solutions, Berlin, Germany
Venguideshe (Venkat) Lakshminarayanan, ServiceNow, Santa Clara, USA
Valli Elangovan, Tata Consultancy Services, Edinburgh, UK
Vijay Shankar, Cognizant, New York, USA
Venkat Sundaram, e-Business International, United States
P. L. Valliappan, ADP Technologies, United States
Venkatesh Srinivason, Tata Consultancy Services Limited, United States
Vasos Pavlika, University College London, UK
Yuvaraj Nagarajan, VIT Infotech, San Ramon, USA
Yamuna Venkates, IStream Jobs, France
Y. D. Sharma, NIT Hamirpur, India
Yegnanarayanan Venkataraman, Kalasalingam Academy of Research and Education, Tamil Nadu, India
Zhiqiang Lin, The Ohio State University, USA

Message from General Chairs

It is really a happy moment for all of us to greet you all at the ICMC 2024, the 10th edition of the highly reputed annual International Conference on Mathematics and Computing. This year, ICMC was organised at the main campus of KALASALINGAM ACADEMY OF RESEARCH AND EDUCATION-KARE located at a small village in the interior of South Tamil Nadu, Krishnankoil-626126, Tamil Nadu, India. The event was held during January 04–07, 2024 followed by a Pre-Conference Symposium on Advanced Mathematical Methods during 02–03, January 2024. It was organised by the Department of Mathematics of School of Advanced Sciences jointly with the School of Computing. We are happy to record the fact that the Ramanujan Mathematical Society (RMS), Cryptology Research Society India (CRSI) and Society for Electronics Transactions and Security (SETS) have participated as joint organisers with KARE. Due to the collective effort of the organisers, we have successfully brought on board 610 eminent personalities as Technical Programme Committee members from all over the world and eighteen eminent speakers from across the globe. It really paved the way for high-quality academic and technical exchange of thought processes between likeminded delegates. As usual, this time also ICMC has created a great impact amongst the participating countries, and everyone have thoroughly enjoyed the academic ambience and the location with the scenic beauty of the southern parts of the Western Ghats of Tamil Nadu. The central theme and topics on mixed areas of Mathematics and Computing created a strong impact. Original research articles published from this 10th edition of ICMC stand as a testimony for the hard work of researchers. We are delighted to record a wonderful fact that a total of 40 papers have been considered for publication in the conference proceeding out of 282 submitted papers through the hard peer-review process by the TPC members. We are much grateful to the following invited speakers for their graceful acceptance and also for having delivered their best of the best speech at the conference. The honourable speakers are Elisa Bertino (Purdue University, USA), Bavani Thuraisingham (University of Texas, USA), Muriel Medard (MIT, USA), Ramamohanarao Kottagiri (University of Melbourne, Australia), Mohammad S. Obaidat (University of Texas-Permian Basin, USA), Sedat Akleylek (Ondokuz Mayis University, Samsun, Turkey), Ekrem Savas (Usak University, Turkey), Bryan

Freyberg (University of Minnesota, USA), Mark Sepanski (Baylor University, Waco Texas, USA), Clare D. Cruz (Chennai Mathematical Institute, India), Jaya Iyer (The Institute of Mathematical Sciences, India), Arvind Ayyar (Indian Institute of Science, India), R. K. Sharma (IIT Delhi, India), Tanmoy Som (IIT (BHU) Varanasi India), Tanmay Basak (IIT Madras, India), Madhumangal Pal (Vidyasagar University, India), A. K. B. Chand (IIT Madras, India) and Vishnu Pendyala (San Jose State University, USA).

We express our sincere gratitude to all the programme and organising committee members for their fantastic review process and other works. We would also like to thank "Illaya Vallal" Dr. Sridharan, Chancellor, and his team at Kalasalingam Academy of Research and Education for their support and excellent infrastructure. **We also thank the sponsors, The Defence Research and Development Organisation (DRDO) and The National Board for Higher Mathematics (NBHM), Government of India, for financial support**. We also appreciate all conference participants for making ICMC a memorable one.

General Chairs

P. K. Saxena, Scientific Consultant—Cyber Security and Former Director, SAG, DRDO, India
P. D. Srivastava, Indian Institute of Technology Bhilai, Raipur, India

Message from Programme Chairs

We are used to the practice of allotting the task of organising the series of ICMC conferences to good institutions spread across the country. This time, we are much pleased to award the task of organising the 10th edition of ICMC 2024 to the KALASALINGAM ACADEMY OF RESEARCH AND EDUCATION-KARE located at a small village in the interior of South Tamil Nadu, Krishnankoil-626126, Tamil Nadu, India. The event was held during January 04–07, 2024, at the KARE University main campus followed by a Pre-Conference Symposium on Advanced Mathematical Methods during January 02–03, 2024. It was organised by the Department of Mathematics of School of Advanced Sciences jointly with the School of Computing. The aim is to provide a common platform for researchers and experts from both Mathematics and Computing to meet, exchange ideas and learn the recent happenings in the respective fields. The speakers are carefully selected to give the best experience to the prospective audience and young researchers. Eighteen speakers from India and abroad delivered their talks, and some of them gracefully accepted to act as session chairs. The response was really wonderful this time to our call for paper through Easychair. We received 282 papers for the conference. There was an overwhelming response from eminent people of top-class institutions from all over the world. We are fortunate to have 610 members in our Technical Programme Committee. We followed triple blind-review process and carefully selected only 40 articles for publication in the conference proceedings published by the Springer series: *Lecture Notes in Networks and Systems.*

The 10th edition of ICMC 2024 has earned a good repute across countries like India, the USA, Canada, Australia, Japan, France, Germany, China, Indonesia, Turkey, the UAE and Nigeria. The delegates from these countries participated and exchanged their thoughts in a variety of areas of pure mathematics, applied mathematics and computing. We are very thankful to the chief patron, patrons, co-patrons, general chairs, programme chairs, organising chair, speakers, participants, referees, organisers, sponsors and funding agencies for their support and help. Our special thanks to Ramanujan Mathematical Society, Cryptology Research Society of India and Society for Electronic Transactions and Security for coming forward to organise this event jointly with us. Last but not the least, we record here our soulful thanks

to all the volunteers and the workers at the grassroots level who worked tirelessly
to make this event a memorable one. A well-planned teamwork was the real reason
behind the success of this conference.

New Jersey, USA	Jaideep Vaidya
Chennai, India	S. Ponnusamy
Columbus, USA	Zhiqiang Lin
Baltimore, USA	Karuna Pande Joshi
Kolkata, India	Debasis Giri

Preface

Mathematics reveals hidden patterns that help us to understand the world around us. Now, much more than arithmetic and geometry, mathematics today is a diverse discipline that deals with data, measurements and observations from science, with inference, deduction and proof; and with mathematical models of natural phenomena, of human behaviour and of social systems. Man is a social animal, and human life depends upon the cooperation of each other. Group work helps social skills. The ability to work together on tasks with others can build various social skills. The importance of Mathematics and Computing as a tool for science and technology is continually increasing. While science and technology have become so pervasive, mathematics and computing education have continued to dominate the curriculum and remain a key subject area requirement in the higher education and the employment sector. Mathematics and Computing are being applied to agriculture, ecology, epidemiology, tumour and cardiac modelling, DNA sequencing and gene technology. They are used to manufacture medical devices and diagnostics, and sensor technology.

The ICMC conference series began its service since 2013 at Haldia Institute of Technology, India. ICMC was further conducted by many reputed institutes such as IIT (BHU), KIIT, Bhubaneswar and Sikkim University, India. The 10th edition of the ICMC series has been conducted in offline mode. ICMC 2024 will provide a wonderful opportunity for both young and seasoned scientists to meet each other to share new ideas and to provide a space for researchers from both academic and industry to present their original work in the area of Computational Applied Mathematics that comprises topics such as Operations Research, Numerical Analysis, Computational Fluid Mechanics, Soft Computing, Cryptology and Security Analysis, Image Processing, Big Data, Cloud Computing, Data Analytics, IoT, Pervasive Computing, Computational Graph Theory and other emerging areas of research.

The 10th Internal Conference on Mathematics and Computing (ICMC 2024) was organised by the Department of Mathematics of School of Advanced Sciences and School of Computing, of Kalasalingam Academy of Research and

Education-KARE, Krishnankoil-626126, Tamil Nadu, India. The Ramanujan Mathematical Society (RMS), Cryptology Research Society India (CRSI) and Society for Electronics Transactions and Security (SETS) also participated as joint organisers of ICMC 2024.

Original research articles published from this 10th edition of ICMC stand as a testimony to the hard work of researchers. We are delighted to record a wonderful fact that a total of 40 papers have been considered for publication in the conference proceeding out of 282 submitted papers through the hard peer-review process by the TPC members. We are much grateful to the following invited speakers for their graceful acceptance and also for having delivered their best of the best speech at the conference. The honourable speakers are Elisa Bertino, Purdue University, USA, Bavani Thuraisingham, University of Texas, USA, Muriel Medard, MIT, USA, Ramamohanarao Kottagiri, University of Melbourne, Australia, Mohammad S. Obaidat, USA, Sedat Akleylek, Ondokuz Mayis University, Samsun, Turkey, Ekrem Savas, Usak University, Turkey, Bryan Freyberg, University of Minnesota, USA, Mark Sepanski, Baylor University, Waco Texas, USA, Clare D. Cruz, CMI, India, Jaya Iyer, IMSC, India, Arvind Ayyar, ISSC, India, R. K. Sharma, IIT Delhi, Tanmoy Som, IIT (BHU), India, Tanmay Basak, IIT Madras, Madhumangal Pal, Vidyasagar University, India, A. K. B. Chand IIT Madras, India and Vishnu Pendyala, San Jose State University, USA.

A unique feature of this book series is that quality assurance is facilitated by a rigorous selection process through the Easychair submission method. Six hundred and ten eminent people from all over the world have participated as TPC members to select the submitted papers. This book contains carefully papers of high-quality researchers in two volumes of 20 papers each as chapters. These two volumes (Volume I and Volume II) speak an exhaustive literature survey, the bottlenecks and advancements in several areas of mathematics and computing that happened in this decade. The excellent coverage of these two volumes is at a higher level to fulfil the global requirements of mathematics, computing and their applications in science and engineering. The audience of this book are mainly researcher scholars, scientists, mathematicians and people from industry.

As Volume Editors of these two volumes, we gratefully acknowledge all the administrative authorities of Kalasalingam Academy of Research and Education-KARE, for their encouragement and support. We also express our appreciation to all the faculty members and research scholars of the Department of Mathematics and School of Computing of Kalasalingam Academy of Research and Education-KARE. We specially thank the Chief Patron, "Illaya vallal" Dr. Sridharan, the Chancellor of KARE, General Chairs, programme chairs and all the members of the organising committee of ICMC 2024 who contributed as one unit by dedicating their time to make the conference a memorable one. We sincerely acknowledge all the referees for their valuable time in reviewing the manuscripts and for carefully picking the best original research articles for publication. We also record our special mention to the sponsors, The Defence Research and Development Organisation (DRDO) and The National Board for Higher Mathematics (NBHM), Government of India, for liberal financial grant. Finally, we are very glad to Springer (Lecture Notes in Networks and Systems) for their encouragement and guidance towards the publication of the proceedings of the conference as two volumes.

Kolkata, India   Debasis Giri
New Jersey, USA   Jaideep Vaidya
Chennai, India   S. Ponnusamy
Columbus, USA   Zhiqiang Lin
Baltimore, USA   Karuna Pande Joshi
Srivilliputhur, India   V. Yegnanarayanan

Contents

Editors and Contributors

About the Editors

Debasis Giri is at present Associate Professor in the Department of Information Technology of Maulana Abul Kalam Azad University of Technology (Formerly known as West Bengal University of Technology), West Bengal, India prior to Professor (in Computer Science and Engineering) and Dean (in School of Electronics, Computer Science and Informatics) of Haldia Institute of Technology, Haldia, India. He did his masters (M.Tech. and M.Sc.) both from IIT Kharagpur, India, and also completed his Ph.D. from IIT Kharagpur, India. He is tenth all India rank holder in Graduate Aptitude Test in Engineering in 1999. He has published more than 100 papers in international journal/conference. His current research interests include Cryptography, Information Security, Blockchain Technology, E-commerce Security and Design and Analysis of Algorithms. He is Editorial Board Member and Reviewer of many International Journals. He is also Program Committee Member of International Conferences. He is a life member of Cryptology Research Society of India, Computer Society of India, the International Society for Analysis, its Applications and Computation (ISAAC) and IEEE annual member.

Jaideep Vaidya is a Distinguished Professor of Computer Information Systems at Rutgers University and the Director of the Rutgers Institute for Data Science, Learning, and Applications. He received the B.E. degree in Computer Engineering from the University of Mumbai, the M.S. and Ph.D. degree in Computer Science from Purdue University. His general area of research is in security, privacy, data mining, and data management. He has published over 200 technical papers in peer-reviewed journals and conference proceedings, and has received several best paper awards from the premier conferences in data mining, databases, digital government, security, and informatics. He is an IEEE and AAAS Fellow as well as an ACM

Distinguished Scientist. He served as the Editor in Chief of the IEEE Transactions on Dependable and Secure Computing.

S. Ponnusamy is currently the Chair Professor at IIT Madras, and the President of the Ramanujan Mathematical Society, India. His research interest includes complex analysis, special functions, and functions spaces. He served five years as a Head of the Indian Statistical Institute, Chennai Centre. He is the Chief Editor of the Journal of Analysis and serves as a Editorial member for many peer reviewed international journals. He has written five text books and has edited several volumes, and international conference proceedings. He has solved several long standing open problems and conjectures, and published more than 300 research articles in reputed international journals. He has been a Visiting Professor to a number of universities in abroad (e.g. Hengyang Normal University, Hunan First Normal University and Hunan Normal University; Kazan Federal University and Petrozavodsk State University; University Sains Malaysia; University of Aalto, University of Turku, and University of Helsinki; University of South Australia; Texas Tech University). Currently, he is also a Leader of the group on the geometric theory of functions at the Laboratory "Multidimensional Approximation and Applications" of the Lomonosov Moscow State University, Moscow Center for Fundamental and Applied Mathematics, Moscow, Russia. He is also a Chair Professor "Furong Scholars Award Program", of Hunan First Normal University, China.

Zhiqiang Lin is a Distinguished Professor of Engineering, and the director of Institute for Cybersecurity and Digital Trust (ICDT) at The Ohio State University. His research interests center around systems and software security, with a key focus on (1) developing automated binary analysis techniques for vulnerability discovery and malware analysis, (2) hardening the systems and software from binary code rewriting, virtualization, and trusted execution environment, and (3) the applications of these techniques in Mobile, IoT, Bluetooth, and Connected and Autonomous Vehicles. He has published over 140 papers, many of which appeared in the top venues in cybersecurity. He is an ACM Distinguished Member, a recipient of Harrison Faculty Award for Excellence in Engineering Education, NSF CAREER award, AFOSR Young Investigator award, and Outstanding Faculty Teaching Award. He received his Ph.D. in Computer Science from Purdue University.

Karuna Pande Joshi is an Associate Professor of Information Systems at the University of Maryland, Baltimore County (UMBC). She is the UMBC Director for the Center of Accelerated Real Time Analytics (CARTA) and the Director of the Knowledge, Analytics, Cognitive, and Cloud (KnACC) Lab. She is also the Undergraduate Program Director of the Business Technology Administration Program. Her primary research focus is Data Science, Legal Text Analytics, Cloud Computing, and Health IT. She has published over 90 technical papers in peer-reviewed journals and conference proceedings. Dr. Joshihas been awarded research grants by NSF, ONR, DoD, NIH, Cisco, and GE Research. She received her M.S. and Ph.D. in Computer Science from UMBC, where she was twice awarded the IBM Ph.D. Fellowship. She did her

Bachelor of Engineering (Computers) from the University of Mumbai. Dr. Joshi has also worked for over 15 years in the Industry, including as a Senior Information Management Officer at the International Monetary Fund for nearly a decade.

V. Yegnanarayanan is a Senior Professor of Mathematics at Kalasalingam Academy of Research and Education, Tamilnadu, India. His Erdos Number is three. He has authored 215 Research papers in reputed refereed journals and Conferences and published eight Patents in India. So far he has produced six Ph.D's. He was elected as a Senior Member of IEEE for his meritorious contributions to the cause of technical education and research in 2012. He has won the prestigious Sentinel of Science Award by Publons, UK in the year 2016 for his contributions to Review work of research papers submitted to journals in Mathematics. He is a life member and affiliate members of various professional societies like AMS, SIAM, RMS, IMS, ISTE. He has successfully completed funded research projects and organized SDP's funded by AICTE, Conferences sponsored by Tamilnadu State Council for Science and Technology and delivered a number of invited talks in India and Abroad.

Contributors

Ali Al-Sharadqah California State University Northridge, Northridge, CA, USA; Prince Mohammad Bin Fahid University, Dhahran, Saudi Arabia

Christopher Samuel Raj Balraj International College of the Cayman Islands, Grand Cayman, Cayman Islands

Rachna Bhatia Department of Mathematics, School of Advanced Sciences, Vellore Institute of Technology, Vellore, Tamilnadu, India

Mainak Chaudhury Crypto Research Lab, IIT Kharagpur, Kharagpur, India

Dipanwita Roy Chowdhury Department of Computer Science and Engineering, Indian Institute of Technology Kharagpur, Kharagpur, West Bengal, India; Crypto Research Lab, IIT Kharagpur, Kharagpur, India

Abhijit Das Crypto Research Lab, IIT Kharagpur, Kharagpur, India

S. Dhanasekar Mathematics, School of Advanced Sciences, Vellore Institute of Technology, Chennai, Tamilnadu, India

Ankita Dhar Department of Computer Science and Engineering, Sister Nivedita University, Newtown, West Bengal, India

Sumathi Ganesan Department of Mathematics, CEG Campus, Anna University, Chennai, India

Shuddhashil Ganguly Department of Computer Science and Engineering, Sister Nivedita University, Newtown, West Bengal, India

Chia-Hsin Huang National Sun Yat-sen University, Kaohsiung, Taiwan

R. Ishwariya Department of Science and Humanities, Amrita School of Engineering, Amrita Vishwa Vidyapeetham, Chennai, Tamil Nadu, India

S. K. Hafizul Islam Department of Computer Science and Engineering, Indian Institute of Information Technology Kalyani, West Bengal, India

J. Jane Rubel Angelina Kalasalingam Academy of Research and Education, Krishnankovil, TN, India

Pratibha Joshi Department of Mathematics, AIAS, Amity University, Noida, India

Arijit Karati National Sun Yat-sen University, Kaohsiung, Taiwan

Aman Kishore Department of Computer Science and Engineering, Indian Institute of Information Technology Kalyani, West Bengal, India

Kundakarla Syam Kumar Kalasalingam Academy of Research and Education, Krishnankovil, TN, India

G. Mahalakshmi Department of Information Science and Technology, CEG Campus, Anna University, Chennai, India

Kalpana Mahalingam Department of Mathematics, Indian Institute of Technology Madras, Chennai, India

Matteo Marciano Gazelien Records Lab, Department of Arts and Humanities, Music Program, New York University Abu Dhabi, Abu Dhabi, United Arab Emirates

Prashanthi Devi Marimuthu Department of Environmental Science and Management, School of Environmental Sciences, Bharathidasan University, Tiruchirappalli, Tamil Nadu, India

Anisha Mitra Department of Computer Science and Engineering, Indian Institute of Technology Kharagpur, Kharagpur, West Bengal, India

Himadri Mukherjee TISA Lab, Department of Computer Science, West Bengal State University, Berunanpukuria, West Bengal, India

P. Nagaraj Kalasalingam Academy of Research and Education, Krishnankoil, Tamilnadu, India

Ola Nusierat Prince Mohammad Bin Fahid University, Dhahran, Saudi Arabia

Debranjan Pal Crypto Research Lab, IIT Kharagpur, Kharagpur, India

Giuliano Piga California State University Northridge, Northridge, CA, USA

Jitesh Pradhan Department of Computer Science and Engineering, National Institute of Technology Jamshedpur, Jharkhand, India

G. Rajalaxmi Department of Data Science, Bishop Heber College, Tiruchirappalli, Tamil Nadu, India

Helda Princy Rajendran Department of Mathematics, Indian Institute of Technology Madras, Chennai, India

Kaushik Roy TISA Lab, Department of Computer Science, West Bengal State University, Berunanpukuria, West Bengal, India

I. Sakthidevi Adhiyamaan College of Engineering, Hosur, TN, India

Amit Sardar Indian Institute of Technology, Kharagpur, India

S. Sasirekha National Institute of Technical Teachers Training and Research, Chennai, India

Janani Selvaraj Department of Data Science, Bishop Heber College, Tiruchirappalli, Tamil Nadu, India

M. Sheela Rani Mathematics, School of Advanced Sciences, Vellore Institute of Technology, Chennai, Tamilnadu, India

Po-An Shih National Sun Yat-sen University, Kaohsiung, Taiwan

Aman P. Singh Department of Computer Science and Engineering, Indian Institute of Information Technology Kalyani, West Bengal, India

S. J. Subhashini SRM Madurai College for Engineering and Technology, Madurai, TN, India

N. Sundareswaran Kalasalingam Academy of Research and Education, Krishnankovil, Tamilnadu, India

Anand Kumar Tiwari Department of Applied Science, Indian Institute of Information Technology, Allahabad, Uttar Pradesh, India

Amit Tripathi Department of Applied Science and Humanities, Rajkiya Engineering College Banda, Atarra, Uttar Pradesh, India

Purnendu Vashistha Department of Computer Science and Engineering, Indian Institute of Information Technology Kalyani, West Bengal, India

M. Vijay Kalasalingam Academy of Research and Education, Krishnankovil, Tamilnadu, India

S. E. Vimal Department of Data Science, Bishop Heber College, Tiruchirappalli, Tamil Nadu, India

K. Vivekrabinson Kalasalingam Academy of Research and Education, Krishnankovil, Tamilnadu, India

Cheng-Che Wu National Sun Yat-sen University, Kaohsiung, Taiwan

Venkataraman Yegnanarayanan Kalasalingam Academy of Research and Education, Krishnankovil, TN, India

Deep Learning-Based Differential Distinguishers for NIST Standard Authenticated Encryption and Permutations

Debranjan Pal, Mainak Chaudhury, Abhijit Das, and Dipanwita Roy Chowdhury

Abstract Deep learning-based cryptanalysis is one of the new ideas that has emerged in recent years. By using deep learning-based methodologies, researchers are currently modeling conventional differential cryptanalysis. We use deep learning models, *CNN, LSTM, LGBM, DenseNet*, and *LeNet*, to generate deep learning-based differential distinguishers that can reveal weaknesses in the encryption schemes. We focus on National Institute of Standards and Technology (NIST) standard lightweight authenticated encryption (AE), such as *TGIF-TBC* and *LIMDOLEN-128*, along with permutation methods like *SPARKLE-256, ACE-128*, and *SPONGENT-160*. Our research has led us to find that deep learning techniques can generate differential distinguishers for these cryptographic elements. Specifically, we were able to develop differential distinguishers for the *SPONGENT-160* permutation up to 7 rounds, for the *SPARKLE-256* permutation up to 3 rounds, for the *ACE-128* permutation up to 4 rounds, for the *TGIF-TBC* AE up to 5 rounds, and for the *LIMDOLEN-128* AE up to 14 rounds. Notably, this marks the first instance of a deep learning-based differential classifier for the authenticated encryptions *TGIF-TBC, LIMDOLEN-128*, as well as the permutations *SPARKLE-256, ACE-128*, and *SPONGENT-160*, based on our current understanding. When considering various models, both *DenseNet* and *CNN* demonstrate strong performance. However, it is the LightGBM (*LGBM*) model that truly shines as the optimal choice, primarily attributed to its minimal parameter requirements and rapid response speed.

Keywords Deep learning · Authenticated encryption · Permutation · Differential cryptanalysis · Neural distinguisher · NIST

D. Pal (✉) · M. Chaudhury · A. Das · D. R. Chowdhury
Crypto Research Lab, IIT Kharagpur, Kharagpur, India
e-mail: debranjanpal@iitkgp.ac.in

A. Das
e-mail: abhij@cse.iitkgp.ac.in

D. R. Chowdhury
e-mail: drc@cse.iitkgp.ac.in

© The Author(s), under exclusive license to Springer Nature Singapore Pte Ltd. 2024
D. Giri et al. (eds.), *Proceedings of the Tenth International Conference on Mathematics and Computing*, Lecture Notes in Networks and Systems 963,
https://doi.org/10.1007/978-981-97-2069-9_1

1 Introduction

In recent days, machine learning and deep learning are dominating the world in computer vision, natural language processing, pattern recognition, health care, and law, to name a few. Researchers have applied machine learning in the cryptology domain for the past few years. In the 1990s, Ronald 1. Rivest [12] presented the similarities of machine learning and cryptography and said about the two domains being "sister fields". The research done since then by numerous researchers has demonstrated that this is accurate. Differential cryptanalysis is one of the most potent and old methods of traditional cryptanalysis. Differential cryptanalysis is a technique for breaking encryption schemes by studying how differences in the input affect the output. It is mainly used for block ciphers and against stream ciphers and hash functions. The idea is to find pairs of plaintexts with a fixed difference (usually XOR) and observe the difference in the corresponding ciphertexts. By doing this, the attacker can identify patterns or probabilities that reveal information about the secret key or the encryption algorithm. In 2019, Gohr [6] in CRYPTO 2019 proposes a deep learning-based ND that performs round-reduced differential cryptanalysis up to 11 rounds and performs key recovery for the same. In 2022, Perusheska et al. [11] perform deep learning-based cryptanalysis of different AES modes of operation. Pal et al. [9, 10] propose differential classifiers for PRIDE, RC5, HIGHT, LEA, and SPARX in 2022. In 2022, Lu et al. also introduce differential NDs for SIMON and SIMECK block ciphers. In 2022 and 2023, Baksi et al. in [2, 3] propose ND-based differential cryptanalysis attacks on many authentication encryption schemes and some permutations. This paper redefines appropriate deep learning models *LSTM*, *LGBM*, *CNN*, and *DenseNet* for searching differential distinguishers. We found distinguishers for 7 rounds of *SPONGENT-160* [5], 3 rounds out of 7 rounds of *SPARKLE-256* permutation [4], 4 rounds out of 16 rounds of *ACE-128* Permutation [1], 5 rounds out of 18 rounds of *TGIF-TBC* [7], and 14 rounds out of 16 rounds of *LIMDOLEN-128* [8] encryption function.

The rest of the paper is organized as follows. In Sect. 2, we present the models for constructing the deep neural classifier, the architecture of the corresponding models. Section 3 explains the implementation details. Section 4 presents the experimental results. A brief discussion on model-wise comparative analysis of distinguishers is provided in Sect. 5. Section 6 concludes the paper.

2 Deep Learning-Based Classifier

Gohr [6] first proposes the principle of distinguishing a cipher from a random text. The input for Gohr's algorithm is the plaintext difference I_Δ. Assume $(\mathbb{P}_1, \mathbb{P}_2)$ be a plaintext pair corresponding to the ciphertext pair $(\mathbb{C}_1, \mathbb{C}_2)$ after encryption of R_n rounds. Create an output label O_L so that for each pair of plaintext and ciphertext, we can deduce

$$\begin{cases} O_L = 0 & \text{if } \mathbb{P}_1 \oplus \mathbb{P}_2 \neq I_\Delta \\ O_L = 1 & \text{if } \mathbb{P}_1 \oplus \mathbb{P}_2 = I_\Delta \end{cases}$$

This primarily aims to separate the actual ciphertext pairs from random ciphertext pairs. As a result, we select the right plaintext difference for a cipher with a preset round for constructing the dataset. Then, we build a deep learning model and train it using the dataset. To create the validation dataset, apply the same plaintext difference to the training dataset and then check the training accuracy to see if it is greater than 50%. A differential classifier is identified for the associated cipher with the specified rounds if the validation accuracy exceeds 50%.

Algorithm 1 Data generation of cipher

Inputs: Number of rounds(R_n), Plaintext difference(I_Δ), Number of data(N_d)
Output: Generated dataset $\mathbb{D}_{T/V}$
  procedure DATAGEN(R_n, I_Δ, N_d)
    for m $\in [1, N_d]$ **do**
      Let $f \leftarrow GenRandom()$ ▷ Generate an one bit random number
      Let $\mathbb{P}_1 \leftarrow GenRandom()$ ▷ Random plaintext
      Let $\mathbb{K} \leftarrow GenRandom()$ ▷ Random Key
      Let $\mathbb{D}_{T/V} \leftarrow \phi$ ▷ Dataset Collection Object
      Let $R_\Delta \leftarrow GenRandom()$ ▷ Random plaintext difference
      if f = 1 **then**
        $\mathbb{P}_2 \leftarrow \mathbb{P}_1 \oplus I_\Delta$
        $\mathbb{C}_1 \leftarrow Cipher\_Encryption(\mathbb{P}_1, \mathbb{K}, R_n)$
        $\mathbb{C}_2 \leftarrow Cipher\_Encryption(\mathbb{P}_2, \mathbb{K}, R_n)$ ▷ Encryption Oracle
        $\mathbb{D}_{T/V} \leftarrow [\mathbb{C}_1, \mathbb{C}_2, \mathbb{C}_1 \oplus \mathbb{C}_2, 1]$
      else
        $\mathbb{P}_2 \leftarrow \mathbb{P}_1 \oplus R_\Delta$
        $\mathbb{C}_1 \leftarrow Cipher\_Encryption(\mathbb{P}_1, \mathbb{K}, R_n)$
        $\mathbb{C}_2 \leftarrow Cipher\_Encryption(\mathbb{P}_2, \mathbb{K}, , R_n)$
        $\mathbb{D}_{T/V} \leftarrow [\mathbb{C}_1, \mathbb{C}_2, \mathbb{C}_1 \oplus \mathbb{C}_2, 0]$
      end if
    end for
    return $\mathbb{D}_{T/V}$
  end procedure

The creation of datasets for training and testing is described in Algorithm 1. The input consists of three components: the plaintext difference (I_Δ), the number of data items (N_D), and the round number (R_n) up to which we want to discover the neural classifier for the specified cipher. Randomly generate one plaintext $\mathbb{P}_1$ and select a second plaintext $\mathbb{P}_2$ ensuring that $I_\delta = \mathbb{P}_1 \oplus \mathbb{P}_2$. Real pair refers to the pair ($\mathbb{P}_1, \mathbb{P}_2$). Pick one random plaintext pair ($\mathbb{P}_1', \mathbb{P}_2'$). Encryption is now carried out using a random oracle with a given round number R_n to obtain the ciphertext pair ($\mathbb{C}_1, \mathbb{C}_2$) from ($\mathbb{P}_1, \mathbb{P}_2$) and ($\mathbb{C}_1', \mathbb{C}_2'$) from ($\mathbb{P}_1', \mathbb{P}_2'$). Store the random ciphertext pair ($\mathbb{C}_1', \mathbb{C}_2'$) with label 0 and the genuine ciphertext pair ($\mathbb{C}_1, \mathbb{C}_2$) with label 1.

Algorithm 2 Algorithm to search a distinguisher

Inputs: Deep learning model MLD_{I_Δ}
Outputs: Training and validation accuracy $(\mathbb{A}_T, \mathbb{A}_V)$

1: **procedure** SEARCHDISTINGUISHER(MLD_{I_Δ})
2:    Load the model MLD_{I_Δ}.
3:    $\mathbb{D}_T \leftarrow$ DataGen(R_n, I_Δ, N_d) ▷ Returns the training data
4:    Train MLD_{I_Δ} applying dataset $\mathbb{D}_T$.
5:    Assume $\mathbb{A}_T$ be the training accuracy
6:    **if** $\mathbb{A}_T > 0.5$ **then**
7:        $\mathbb{D}_V \leftarrow$ DataGen(R_n, I_Δ, N_d) ▷ Returns the validation data
8:        Load the pretrained model MLD_{I_Δ}.
9:        Perform validation testing for MLD_{I_Δ}.
10:       Let $\mathbb{A}_V$ be the validation accuracy.
11:       **if** $\mathbb{A}_V > 0.5$ **then**
12:           Return $(\mathbb{A}_T, \mathbb{A}_V)$
13:       **else**
14:           Return "Distinguisher does not exists"
15:       **end if**
16:       Return $\mathbb{A}_V$
17:   **else**
18:       Return "Distinguisher does not exists"
19:   **end if**
20: **end procedure**

2.1 Architecture Stack of Deep Neural Models

The architecture stack of a deep neural model represents layers inside the model used. It shows the input's dimensions and the input's modified dimensions after it goes through several layers depending on the model chosen and the number of layers used. Using the stack, we find the parameters in each layer used. For example, the layers conv, lstm, batch normalization, etc. use parameters, whereas layers pool, concatenate, and activation function ReLU have no parameters to deal with. Also, there are two types of parameters, trainable and non-trainable parameters. These parameters are trained/untrained during the model's training with the training dataset and can be changed explicitly to what parameters to train and not train. Libraries like TensorFlow and Keras set the trainability of parameters. For *CNN, LSTM, DenseNet,* and *LeNet*, the architecture stacks are shown in Figs. 1a, b and 2a, b, respectively.

3 Implementation Details

We have implemented the data generation algorithms on Intel i7-7700HQ @ 2.8GHz, 4GB NVIDIA GTX 1050Ti on Nobara 36 Workstation. Python3 is used for writing code and utilizes various machine learning libraries such as TensorFlow, Keras, and Microsoft's *LGBM* library. The dataset created is split into a 50% training set and a

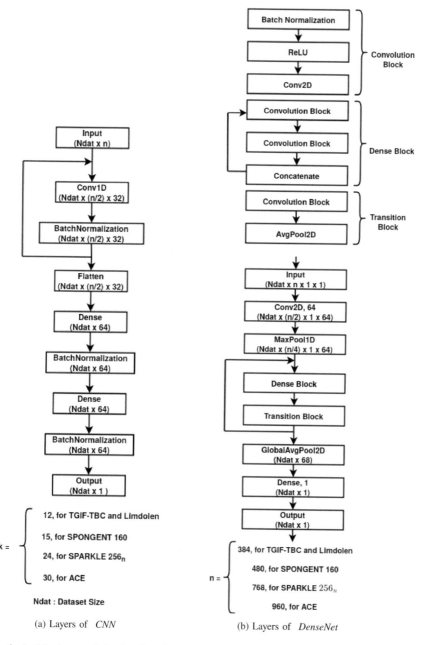

(a) Layers of *CNN*   (b) Layers of *DenseNet*

Fig. 1 Architecture stack for deep learning models *CNN* and *DenseNet*

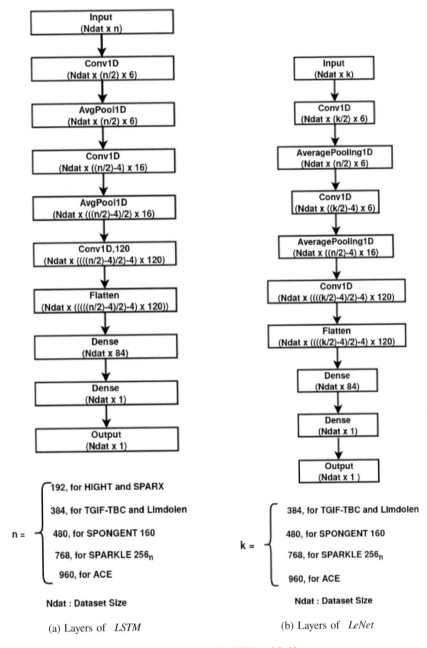

(a) Layers of *LSTM*

(b) Layers of *LeNet*

Fig. 2 Architecture stack for deep learning models *LSTM* and *LeNet*

50% testing set. The deep neural models are trained for ten epochs, and a batch size of 300 is used during the training process.

4   Results

In this section, we will discuss the achieved results on the different distinguishers for all the ciphers and permutations in terms of training accuracy (A_T), validation accuracy (A_V), true positive rate (*TPR*), and true negative rate (*TNR*).

4.1   SPONGENT-160

For *SPONGENT-160*, we choose a plaintext difference where the last bit is one, i.e., $0x0/0x0/0x0/0x1$. The results of *SPONGENT-160* for *CNN, LGBM, LSTM, DenseNet*, and *LeNet* are given in Table 1. The *CNN* and *LGBM* models exhibit high accuracy in the initial rounds, but experience a decline in validation accuracy as the rounds progress. The *LSTM* model shows consistently low accuracy throughout the rounds. The *DenseNet* model achieves high accuracy initially, but experiences a notable drop in validation accuracy in the later rounds. The *LeNet* model demonstrates high accuracy initially, but also experiences a decline in accuracy in the later rounds (Table 2).

Table 1 Training and validation statistics of NDs for *SPONGENT-160*

Models	R_n	A_T	A_V	TPR	TNR
CNN	5	100.00	99.99	1.000	0.999
	6	99.55	70.37	0.712	0.720
	7	99.26	50.36	0.501	0.505
LGBM	5	99.00	98.82	0.981	0.991
	6	81.72	78.95	0.754	0.876
	7	77.62	50.62	0.502	0.503
LSTM	5	50.68	49.70	0.497	0.431
	6	50.00	50.25	0.500	0.501
	7	49.98	50.26	0.501	0.504
DenseNet	5	99.08	96.52	0.965	0.967
	6	78.50	68.96	0.638	0.792
	7	63.26	50.36	0.503	0.500
LeNet	5	99.95	99.87	0.998	0.997
	6	67.30	67.15	0.653	0.714
	7	52.70	50.33	0.501	0.503

Table 2 Training and validation statistics of NDs for *ACE-128*

Models	R_n	A_T	A_V	TPR	TNR
CNN	2	100.00	100.00	1.000	1.000
	3	100.00	99.84	1.000	0.997
	4	100.00	50.03	0.501	0.499
LGBM	2	100.00	100.00	1.000	1.000
	3	99.32	98.65	1.000	0.974
	4	79.84	49.97	0.495	0.499
LSTM	2	82.39	82.24	0.838	0.806
	3	53.73	53.15	0.534	0.536
	4	50.17	49.83	1.000	0.498
DenseNet	2	100.00	100.00	1.000	1.000
	3	60.77	50.46	0.502	0.507
	4	61.36	49.87	0.503	0.498
LeNet	2	100.00	100.00	1.000	1.000
	3	100.00	99.95	1.000	0.999
	4	53.83	49.94	0.499	0.502

Table 3 Training and validation statistics of the ND for *SPARKLE-256*

Models	R_n	A_T	A_V	*TPR*	*TNR*
cnn	2	100.00	100.00	1.000	1.000
	3	99.98	49.79	0.495	0.499
LGBM	2	100.00	100.00	1.000	1.000
	3	79.28	50.20	0.501	0.502
LSTM	2	79.67	81.08	0.812	0.807
	3	50.11	50.19	0.513	0.504
DenseNet	2	100.00	100.00	1.000	1.000
	3	52.70	50.53	0.503	0.505
LeNet	2	100.00	100.00	1.000	1.000
	3	96.00	96.82	0.957	0.980

4.2 SPARKLE-256

The results presented in Table 3 demonstrate the effects of a plaintext difference where the last bit is set to one, i.e., $0x0/0x0/0x0/0x1$, when using the *SPARKLE-256* algorithm on various deep neural models including *CNN, LGBM, LSTM, DenseNet,* and *LeNet*. The *CNN* model achieves perfect accuracy in the second round but experiences a significant drop in validation accuracy in the third round. The *LGBM* model achieves high accuracy in the second round but also experiences a decline in accuracy in the third round, with the validation accuracy slightly above 50%.

The *LSTM* model shows moderate accuracy, with a slight decrease in accuracy in the third round. The *DenseNet* model initially achieves perfect accuracy but experiences a decline in validation accuracy in the later rounds. The *LeNet* model demonstrates consistently high accuracy in both training and validation, with a slight drop in validation accuracy in the third round.

4.3   ACE-128

The results of *ACE-128* on *CNN*, *LGBM*, *LSTM*, *DenseNet*, and *LeNet* are provided in Table 2 after applying plaintext difference $0x0/0x0/0x0/0x0/0x0/0x0/0x0/0x0/0x0/0x1$. The *CNN* and *LGBM* models initially exhibit high accuracy but experience a notable decrease in validation accuracy in later rounds. The *LSTM* model shows moderate accuracy throughout the rounds, while *DenseNet* achieves high accuracy initially but experiences a decline in validation accuracy. The *LeNet* model consistently maintains high accuracy, with only a slight decrease in validation accuracy in the final round.

4.4   TGIF-TBC

In order to identify a distinguisher with high accuracy, we choose the plaintext difference as $0x80000000/0x00000000/0x00000000/0x00000000$. We then applied this difference to the *CNN*, *LGBM*, *LSTM*, *DenseNet*, and *LeNet* models. The resulting *TGIF-TBC* outcomes for each distinguisher are presented in Table 4. The *CNN* model achieves high accuracy in the earlier rounds but experiences a decrease in both training and validation accuracy in the later rounds. The *LGBM* model demonstrates high accuracy in the first two rounds, but the validation accuracy declines in the subsequent rounds. The *LSTM* model shows moderate accuracy, with a decline in both training and validation accuracy as the number of rounds increases. The *DenseNet* model achieves high accuracy in the first two rounds but experiences a decline in accuracy in the later rounds. The *LeNet* model exhibits consistent accuracy across the rounds, with a slight decrease in both training and validation accuracy in the later rounds.

4.5   LIMDOLEN-128

We select the plaintext difference as $0x0/0x0/0x0/0x40$ for finding a distinguisher with good accuracy. The distinguishers for *LIMDOLEN-128* after applying on *CNN*, *LGBM*, *LSTM*, *DenseNet*, and *LeNet* are provided in Table 5. In the initial rounds, the *CNN* model demonstrates a high level of accuracy, but its accuracy slightly decreases

Table 4 Training and validation statistics of the ND for *TGIF-TBC*

Models	R_n	A_T	A_V	TPR	TNR
CNN	3	99.71	77.52	0.763	0.788
	4	99.47	53.52	0.532	0.538
	5	99.58	49.77	0.502	0.496
LGBM	3	94.48	93.68	0.915	0.961
	4	69.99	61.91	0.615	0.991
	5	77.16	50.08	0.499	0.501
LSTM	3	55.25	53.39	0.552	0.524
	4	50.32	50.17	0.500	1.000
	5	50.19	49.95	0.499	0.499
DenseNet	3	90.57	90.58	0.929	0.945
	4	77.76	55.25	0.822	0.882
	5	61.29	49.77	0.499	0.499
LeNet	3	81.04	80.56	0.788	0.825
	4	59.49	58.60	0.601	0.563
	5	52.15	50.12	0.501	0.502

Table 5 Training and validation statistics of the ND for *LIMDOLEN-128*

Model	R_n	A_T	A_V	TPR	TNR
CNN	12	94.78	51.41	0.516	0.513
	13	98.20	51.72	0.516	0.518
	14	99.41	50.29	0.501	0.503
LGBM	12	73.73	54.00	0.531	0.558
	13	75.80	52.01	0.517	0.523
	14	75.88	51.81	0.514	0.522
LSTM	12	51.10	50.37	0.497	0.558
	13	50.70	50.51	0.506	0.506
	14	50.70	50.11	0.501	0.499
DenseNet	12	62.51	50.78	0.508	0.507
	13	61.87	51.40	0.516	0.513
	14	62.27	50.59	0.506	0.506
LeNet	12	54.44	53.14	0.536	0.523
	13	53.79	52.20	0.513	0.528
	14	52.74	51.50	0.510	0.521

as the number of rounds increases. On the other hand, the *LGBM* model maintains relatively stable accuracy across all rounds, although there is a minor decline in validation accuracy. The *LSTM* model consistently performs poorly throughout the rounds, without any notable improvement. The *DenseNet* model achieves moderate accuracy, but it experiences a slight decrease in both training and validation accuracy in the later rounds. Similarly, the *LeNet* model follows a similar pattern to *DenseNet*, with a slight decline in accuracy as the rounds progress.

5 Model-Wise Comparative Analysis of NDs

This section describes the multiple facets of the model for a particular round of reduced cipher/permutation. We choose training time, number of parameters, architecture, and the validation accuracy obtained (see Table 6). For *LGBM*, we use the default parameters obtained from the respective libraries and because of that *LGBM* takes minimum training time than any other models used. Regarding training time, *CNN* takes the maximum time among all others because our models are sequential and consist of more number of layers than any other models we have used. *DenseNet* requires fewer parameters compared to the rest of the deep neural models except *LGBM*. For six-round *SPONGENT-160 LGBM* performs better than all the other models and *LSTM* is the worst performing model. With four rounds of *TGIF-TBC LGBM* is best in this case and *LSTM* performs worst. Also, *LeNet* performs better than *CNN*, *LSTM*, and *DenseNet*. *LGBM*erforms better for 12 rounds of *LIMDOLEN-128*. For three rounds of *ACE-128*, *CNN* performs better, but the number of parameters using worst. The required training time for *CNN* mode is also more compared to others. Concerning overall performance, *LGBM* serves better than others.

6 Conclusion

In our study, we have identified deep learning generated distinguishers that yield clear outcomes. Specifically, we achieved results for 7 rounds out of 80 rounds for the *SPONGENT-160* permutation, 3 rounds out of 7 for the *SPARKLE-256* permutation, 4 rounds out of 16 for the *ACE-128* permutation, 5 rounds out of 18 for *TGIF-TBC*, and 14 rounds out of 16 for the *LIMDOLEN-128* encryption function. Upon analyzing different models, both *DenseNet* and *CNN* demonstrated commendable performance. However, the *LGBM* model stood out as the most advantageous due to its streamlined parameter configuration and rapid processing time. While each distinguisher exhibited strength with certain ciphers and permutations, it was evident that *CNN* and *DenseNet* outperformed *LGBM* and *LeNet*. Conversely, *LSTM* delivered the least favorable results. An in-depth exploration of the factors influencing performance discrepancies can guide future research directions.

Table 6 Model parameters and performance for different ciphers

Cipher	Network	Architecture	Activation function	No. of param.	Train time (sec)	A_V
SPONGENT-160	CNN	32, (32, 32)× 12, 64, 64, 1	ReLU/sigmoid	1,065,601	694	70.37
	LGBM	Default	Sigmoid	default	5	78.95
	LSTM	64, 96, 128, 192, 1	Sigmoid	444,225	54	50.25
	DenseNet	64, 6, 12, 24, 16, 1	ReLU/sigmoid	189,745	48	68.76
	Lenet	6, 16, 120, 84, 1	tanh/sigmoid	1,159,541	22	67.15
SPARKLE-256	CNN	32, (32, 32)× 12, 64, 64, 1	ReLU/ sigmoid	1,655,425	1200	100.00
	LGBM	Default	Sigmoid	Default	5	100.00
	LSTM	64, 96, 128, 192, 1	Sigmoid	1,035,620	54	81.00
	DenseNet	64, 6, 12, 24, 16, 1	ReLU/sigmoid	189,745	48	100.00
	Lenet	6, 16, 120, 84, 1	tanh/sigmoid	1,885,301	52	100.00
ACE-128	CNN	32, (32, 32)× 12, 64, 64, 1	ReLU/sigmoid	2,048,641	1332	99.84
	LGBM	Default	Sigmoid	Default	5	98.65
	LSTM	64, 96, 128, 192, 1	Sigmoid	448,065	54	53.15
	DenseNet	64, 6, 12, 24, 16, 1	ReLU/sigmoid	189,745	48	50.46
	LeNet	6, 16, 120, 84, 1	tanh/sigmoid	2,369,141	22	99.95
TGIF-TBC	CNN	32,(32, 32)× 12, 64, 64, 1	ReLU/ sigmoid	868,993	309	53.52
	LGBM	Default	Sigmoid	default	5	61.91
	LSTM	64, 96, 128, 192, 1	Sigmoid	443,457	54	50.17
	DenseNet	64, 6, 12, 24, 16, 1	ReLU/sigmoid	189,745	48	55.67
	Lenet	6, 16, 120, 84, 1	tanh/sigmoid	917,621	22	58.60
LIMDOLEN-128	CNN	32,(32,32)× 12, 64, 64, 1	ReLU/ sigmoid	868,993	600	51.41
	LGBM	Default	Sigmoid	Default	5	54.00
	LSTM	64, 96, 128, 192, 1	Sigmoid	443,457	73	50.37
	DenseNet	64, 6, 12, 24, 16, 1	ReLU/sigmoid	189,745	48	50.78
	Lenet	6, 16, 120, 84, 1	tanh/sigmoid	917,621	22	53.14

References

1. Aagaard M, AlTawy R, Gong G, Mandal K, Rohit R (2019) Ace: an authenticated encryption and hash algorithm. Submission to NIST LwC Standardization Process
2. Baksi A, Breier J, Chen Y, Dong X (2021) Machine learning assisted differential distinguishers for lightweight ciphers. In: Design, automation and test in Europe conference and exhibition, DATE 2021, Grenoble, France, 1–5 Feb 2021, pp 176–181
3. Baksi A, Breier J, Dasu VA, Hou X, Kim H, Seo H (2023) New results on machine learning-based distinguishers. IEEE Access 11:54175–54187
4. Beierle C, Biryukov A, Cardoso dos Santos L, Großschädl J, Perrin L, Udovenko A, Velichkov V, Wang Q (2020) Lightweight AEAD and hashing using the sparkle permutation family. IACR Trans Symmetric Cryptol (S1):208–261
5. Beyne T, Chen YL, Dobraunig C, Mennink B (2021) Elephant v2.0. Submission to NIST LwC Standardization Process
6. Gohr A (2019) Improving attacks on round-reduced speck32/64 using deep learning. In: Advances in cryptology—CRYPTO, Santa Barbara, CA, USA, 18–22 Aug 2019. Proceedings of the Part II LNCS, vol 11693, pp 150–179
7. Iwata T, Khairallah M, Minematsu K, Peyrin T, Sasaki Y, Sim SM, Sun L (2019) Thank goodness it's friday (tgif). Submission to NIST LwC Standardization Process (Round 1)
8. Mehner CE (2019) Limdolen : a lightweight authenticated encryption algorithm. Submission to NIST LwC Standardization Process
9. Pal D, Mandal U, Chaudhury M, Das A, Chowdhury DR (2022) A deep neural differential distinguisher for ARX based block cipher. IACR Cryptol ePrint Arch:1195
10. Pal D, Mandal U, Das A, Chowdhury DR (2022) Deep learning based differential classifier of PRIDE and RC5. In: Applications and techniques in information security (ATIS 2022), Manipal, India, 30–31 Dec 2022. CCIS, vol 1804. Springer, pp 46–58
11. Perusheska MG, Trpceska HM, Dimitrova V (2022) Deep learning-based cryptanalysis of different AES modes of operation. In: Arai K (ed) Advances in information and communication. Springer International Publishing, Cham, pp 675–693
12. Rivest RL (1991) Cryptography and machine learning. In: Imai H, Rivest RL, Matsumoto T (eds) Advances in cryptology (ASIACRYPT '91), Fujiyoshida, Japan, 11–14 Nov 1991. LNCS Proceedings, vol 739. Springer, pp 427–439

Prediction of Mental Health Issues and Challenges Using Hybrid Machine and Deep Learning Techniques

Christopher Samuel Raj Balraj and P. Nagaraj

Abstract Mental health issues like melancholy, anxiety, and a lack of sleep in young children, teenagers, and adults are the root causes of emotional stress. It influences a person's feelings, thoughts, or reactions to a particular event or scenario. Being in good physical and mental health is a prerequisite for productive work and realizing one's full potential. From childhood to maturity, maintaining one's mental health is crucial. The various causes of mental health concerns that lead to mental illness include stress, social anxiety, depression, obsessive–compulsive disorder, substance addiction, employment issues, and personality disorders. We used openly accessible web datasets to collect the data. The data was label-encoded to improve prediction. The methods employed include logistic regression, Nave Bayes, decision trees, neural networks, and support vector machines. The Decision Tree, the Support Vector Machine, and the neural network, in that order, are the most trustworthy models for stress, depression, and anxiety. The data is put through several machine-learning techniques to produce labels. Based on these classified categories, a model will be created to forecast the mental state of an individual. People over 18 who are working class make up our primary market after finishing, based on the information a user submitted on the website.

Keywords Mental health · Prediction · Machine learning · Deep learning · Hybrid learning · Web page

C. S. R. Balraj (✉)
International College of the Cayman Islands, Grand Cayman, Cayman Islands
e-mail: christopher.rajblraj@icci.edu.ky

P. Nagaraj
Kalasalingam Academy of Research and Education, Anand Nagar, Krishnankoil 626126, Tamilnadu, India
e-mail: nagaraj.p@klu.ac.in

© The Author(s), under exclusive license to Springer Nature Singapore Pte Ltd. 2024
D. Giri et al. (eds.), *Proceedings of the Tenth International Conference on Mathematics and Computing*, Lecture Notes in Networks and Systems 963,
https://doi.org/10.1007/978-981-97-2069-9_2

1 Introduction

The innovative application of hybrid learning methods in the realm of mental health assessment is essential in the field of medical healthcare. In an era characterized by burgeoning technological advancements and an increasing recognition of mental health concerns, this study seeks to bridge the gap between traditional diagnostic approaches and cutting-edge computational techniques [1]. Through amalgamating a wide range of learning algorithms, such as deep learning, ensemble learning, and other advanced models, this study aims to develop a comprehensive predictive framework that can identify different mental health conditions and challenges. This hybrid approach offers a more nuanced and accurate assessment, potentially leading to more effective interventions and personalized treatment plans. This study explores the ethical issues of using cutting-edge technologies in mental health assessments, going beyond the technical aspects. It seeks to find a middle ground between optimizing computational techniques and protecting the autonomy, privacy, and well-being of those evaluated [2].

1.1 Overview of Mental Health Issues and Challenges

It encompasses many conditions that affect a person's emotional, psychological, and social well-being. These conditions can significantly impact an individual's thoughts, feelings, behavior, and overall quality of life. Here is an overview of some common mental health issues and the challenges they pose like Depression, Anxiety Disorders, Bipolar Disorder, Schizophrenia, Eating Disorders, Post-Traumatic Stress Disorder (PTSD), Substance Use Disorders, Attention-Deficit/ Hyperactivity Disorder (ADHD), Personality Disorders, Neurodevelopmental Disorders. Depression is typified by enduring melancholy and despair as well as a loss of interest in or enjoyment from activities. It can lead to various physical and emotional symptoms and impair daily functioning. This category includes conditions such as generalized anxiety disorder, panic disorder, social anxiety disorder, and specific phobias [3]. Anxiety disorders involve excessive worry, fear, or apprehension that can interfere with daily life. Bipolar disorder involves extreme mood swings, cycling between periods of mania (elevated mood, high energy) and depression (low mood, lethargy). It can lead to significant disruptions in daily functioning. Schizophrenia is a severe mental disorder characterized by disturbances in thinking, emotions, and behavior [4]. It may include hallucinations, delusions, disorganized thinking, and impaired social functioning. These involve problematic use of drugs or alcohol that can lead to physical, psychological, and social consequences. Substance use disorders can co-occur with other mental health conditions. These conditions involve enduring patterns of behavior, cognition, and inner experience that deviate markedly from cultural expectations. Examples include borderline, narcissistic, and antisocial personality disorders [5].

1.2   Process for Identification of Mental Health

Both a person's physical and mental health are taken into account when conducting health research. Both are necessary for an individual to lead a happy and healthy life. A person's productivity, ability to think clearly, and even physical health are all impacted by their mental and physical health [6]. One of the most prevalent diseases worldwide is mental illness. Since there is no specific cause for mental illness, it is imperative to exercise caution and vigilance when it comes to one's mental well-being. Mental illnesses include anxiety, bipolar disorder, depression, eating disorders, and stress. As serious as they seem, if a person has them, they are very dangerous. Because of this, we need to treat our mental health with the same consideration and care that we do our physical health [7–10]. Figure 1 represents the flow work of the mental health issues identification process.

Fig. 1 Process of mental health care identification and action

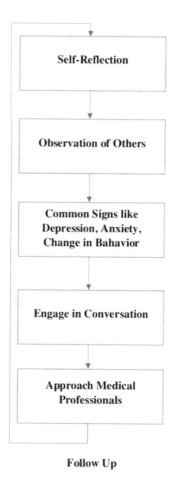

Figure 1 describes how identifying mental health concerns involves a combination of self-awareness, observation of others, and professional assessment. The first thing is to pay attention to your own emotions, thoughts, and behaviors. It recognizes any significant changes or patterns indicating a potential mental health concern. And need to observe the changes in behavior, mood, or habits in friends, family members, or colleagues. After observation, pay attention to identify and recognize the common signs and engage with medical experts to resolve the issues and challenges.

2   Review of Literature Work

To better understand the methods employed in the forecasting of mental health issues, a thorough examination of the literature was carried out. After that, the operation of the algorithms was thoroughly understood. The development of any project must include a literature review. The outcomes of our review of the literature are as follows.

Ansari et al. [11] propose a novel approach that combines multiple machine learning models, known as an ensemble method, to enhance the accuracy and reliability of depression detection. By integrating different algorithms and leveraging their strengths, the ensemble method aims to improve the overall performance of automated depression detection systems. The research likely involves data collection from individuals exhibiting both depressive and non-depressive traits and employs various features and classifiers to develop a comprehensive model. The findings of this study hold promise for more effective and reliable automated tools for identifying depression, which could have significant implications for mental health screening and intervention.

Banna et al. [12] analyzed a unique approach that combines different deep learning models, resulting in a hybrid model. This hybrid model is designed to capture and analyze various textual and contextual information from social media posts. By integrating various aspects of the data, the model aims to provide accurate and timely predictions regarding the impact of the pandemic on mental well-being. The research likely involves the collection and processing of a large volume of social media data to train and validate the hybrid model. The findings of this study have the potential to offer valuable insights into the mental health consequences of public health crises like COVID-19, facilitating more targeted interventions and support strategies.

Mohamed et al. [13] focus on leveraging advanced deep-learning techniques to forecast the effects of the COVID-19 pandemic on mental health using extensive data from social media platforms. The study proposes a unique approach that combines different deep learning models, resulting in a hybrid model. This hybrid model is designed to capture and analyze various textual and contextual information from social media posts. By integrating various aspects of the data, the model aims to provide accurate and timely predictions regarding the impact of the pandemic on mental well-being. The research likely involves the collection and processing of a large volume of social media data to train and validate the hybrid model. The findings of this study have the potential to offer valuable insights into the mental health consequences of public health crises like COVID-19, facilitating more targeted interventions and support strategies.

Ahmad et al. [14] employ a technique called decision-level fusion, which combines the outputs of multiple classifiers. The classifiers used in this research are likely a combination of different machine learning algorithms. By blending their predictions, the hybrid classifier aims to enhance the accuracy and reliability of mental health classification. The research likely involves gathering diverse data on mental health symptoms and conditions. This data is then used to train and evaluate the hybrid classifier, allowing it to categorize mental health disorders effectively. The findings of this study offer potential advancements in the precision of mental health assessments, which could lead to more targeted and effective treatments for individuals dealing with mental health challenges.

Shahid et al. [15] involve implementing a program or intervention to promote self-compassion among students. It then assesses its effects on three key areas: academic motivation, academic stress, and overall mental health. This research is particularly relevant in the hybrid learning environment, which presents unique student challenges. The findings of this study may provide valuable insights into how cultivating self-compassion can potentially improve students' academic experience and mental well-being during such learning modalities.

3   Proposed Methodology

Figure 2 shows the proposed work's architecture for predicting mental health issues and various challenges. The basic outline of the architecture, along with potential challenges, are working as follows. First, gather data from two sources: surveys and electronic health record formats. Then, we performed the data clean, normalized and pre-processed the data to remove noise and ensure consistency. From the collected data, we need to identify the indicative mental health status and relevant features. At last, we need to apply various learning to predict the model based on the nature of the data. Then, divide the dataset into training, validation, and test sets to train and evaluate the models. After the evaluation process, deploy the model with web page integration.

3.1   Dataset Collection

The dataset used in this research is taken from the Kaggle [16]. The dataset contains essential features that can lead to mental health problems. This dataset contains 154 rows and 33 columns. It can predict the accuracy of mental health problems.

C. S. R. Balraj and P. Nagaraj

Fig. 2 Architecture of the proposed mental health issues prediction model

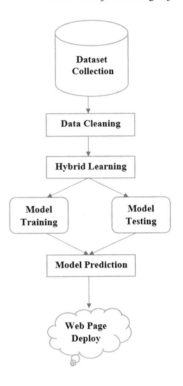

3.2 Data Cleaning

After data exploration, data cleaning will be carried out to remove bias and help build more precise machine-learning models. Our data will then be compiled, and the number of rows and columns of information will be decreased. Data cleaning has the advantage of virtually eliminating long-term unneeded and undesired data while providing accurate input data. Finally, all missing data have been eliminated to eliminate errors made during the model construction.

3.3 Learning Models

The data must be split into training and testing data as the first step in building the model. This model is used to assess the correctness of this model using machine learning techniques. Some of these have also employed machine learning and deep learning models. The models with the highest precision were the hybrid and combined, in addition to the CNN algorithm.

3.3.1 Logistic Regression

A well-known supervised learning technique, logistic regression is one of the most common machine learning techniques. In this strategy, we predict an individual dependent variable from a group of unbiased factors. To forecast the results of structured factors, we employ logistic regression. As a result, the result should be a discrete or categorical value. It may yield probabilistic numbers between 0 and 1 rather than exact numbers like 0 and 1, such as 0 or 1, Yes or No, true or false, etc.

3.3.2 K Nearest Neighbor

A fundamental machine learning algorithm based on the Supervised Learning method is the K-nearest neighbor algorithm. In the K-NN approach, new cases and old cases/data will be comparable. KNN is a non-parametric algorithm that makes no assumptions about the distribution or the highlighted data. Additionally, it works with several classes.

3.3.3 Random Forest

With the help of the supervised machine learning method, the technique known as random forest may solve classification and regression issues. But it is frequently employed in classification. The reason it combines numerous decision trees into a "forest" and feeds random characteristics from the data set being used to them is why it is called a random forest.

3.3.4 Decision Tree

Although Decision Tree is a supervised learning technique that may be applied to classification and regression issues, classification challenges are where it is most frequently applied. A tree-structured classifier has core nodes that reflect dataset attributes, branches that represent decision rules, and leaf nodes that represent the outcome.

3.3.5 Support Vector Machine

Support Vector Machine, or SVM, is a well-known Supervised Learning method for classification and regression problems. However, it is typically utilized in machine learning to address problems with classification.

3.3.6 Naïve Bayes

The Bayes theorem is used by the Naive Bayes, a supervised learning method, to address classification problems. It primarily works with a sizable training dataset for text classification. The Nave Bayes Classifier (NBC) is a straightforward and efficient classification technique that supports the creation of quick machine-learning models with the capacity for rapid prediction. Because it is a classifier that uses probabilities, it makes predictions based on how likely an object is to occur.

3.3.7 Convolutional Neural Network

The Deep Learning neural network architecture known as a Convolutional Neural Network (CNN) is frequently used in computer vision. A computer can understand and interpret photographs or other visual data because of the study of computer vision, a branch of artificial intelligence. A more sophisticated artificial neural network (ANN) called a convolutional neural network (CNN) is frequently used to extract features from datasets using grid-like matrixes.

3.4 Model Prediction

After applying the learning algorithm, the scientist needs to perform training and testing of the model with 75% of data and 25% of data, respectively. Once trained, the model is ready for prediction. For validation, use techniques like k-fold cross-validation to ensure the robustness of the models. After receiving the user's values, the model is loaded, and prediction is performed.

3.5 Model Deployment

After building the model, the user must implement a mechanism to monitor and update predictions as new data becomes available continuously. It establishes a threshold for triggering alerts or interventions when potential mental health issues are detected. The accuracy and relevance of the model's output are validated with clinical expertise once Combined machine-generated predictions are made. Now, the model is ready to deploy in web-based cloud internet applications.

4 Results and Discussions

In this work, we have used Google Collaborators to implement the Python framework for mental health issue prediction. The model seeks to categorize, forecast, and present information visually for the user to see and comprehend. Additionally, it gives the user an interface for simple navigation. All navigations are described in Figs. 4, 5 and 6.

Despite being very sophisticated, machine learning algorithms, also referred to as "black box" algorithms, do not allow us to grasp the full potential of AI. Conversely, hybrid learning techniques are advantageous. They comprise various little algorithms that work in unison to accomplish the most extraordinary and potent task imaginable. Combining two machine learning algorithms or one deep learning algorithm is also possible, lowering uncertainty and delivering more precise results. Without explicit human intervention, computers and other devices can autonomously learn from the past and improve over time due to a type of artificial intelligence known as machine learning. The foundation of machine learning is the creation of computer programs with autonomous data collection and information-gathering capabilities. This is especially helpful in the data-rich healthcare sector because, when adequately fed to an intelligent machine and trained well, the forecast model's output will be unparalleled, error-free, and reduce the time required for diagnosis.

The models with the highest precision were the hybrid and combined, in addition to the CNN algorithm. The result of the predictor model is analyzed in Table 1.

From the above Table 1, we analyzed that the CNN model produces higher accuracy than the other learning models. It is also described in Fig. 3 as a better comparative analysis.

The Higher Accuracy CNN model summary is given below (Figs. 4, 5 and 6).

Table 1 Algorithm analysis results

S. no.	Algorithm name	Accuracy score (%)
1	Convolutional neural network	98.181820
2	Logistic regression	93.750000
3	Random forest	87.500000
4	K-nearest neighbors	83.333333
5	Support vector machine	83.333333
6	Decision tree	266.666667
7	Naive bayes	58.333333

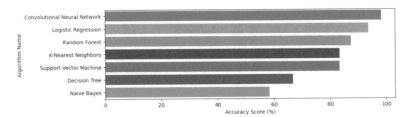

Fig. 3 Comparative analysis of mental health issue predictor model

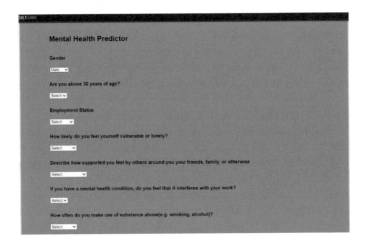

Fig. 4 Mental health predictor attribute value

Fig. 5 Predictor result for healthy object

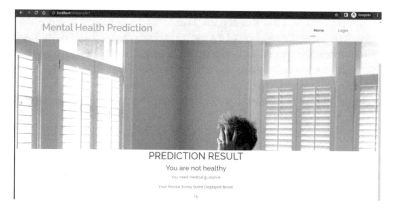

Fig. 6 Predictor result for unhealthy object with score

```
Model: "model"
```

Layer (type)	Output Shape	Param #
input_1 (InputLayer)	[(None, 60)]	0
dense (Dense)	(None, 64)	3904
dense_1 (Dense)	(None, 64)	4160
dense_2 (Dense)	(None, 1)	65

```
Total params: 8,129
Trainable params: 8,129
Non-trainable params: 0
```

5   Conclusion

Numerous methods and algorithms have been developed and tested to assess and treat mental health issues. There are still many problems that can be solved better. Moreover, there are still a lot of issues in the field of machine learning for mental health that need to be identified and investigated in several contexts. Classifying mental health data is typically a challenging task, so the features that machine learning algorithms use will have a significant impact on the classification's performance. So, through this work, we can learn more about mental health, how it affects a person's health, its various forms, how important it is to diagnose and treat it, and how can people with mental illness be detected using hybrid algorithms. Currently, mental health is a sensitive and vital subject. For one to lead a balanced and healthy life, it is essential. The mental condition of an individual affects their thoughts, deeds, and emotions. It might have an impact on a person's productivity and performance. This algorithm will be tested after training, and a model will be created with the highest accuracy of 98% with the CNN model. After that, it would be made available online so that users could fill out forms and get the results when public users wanted.

References

1. Chung J, Teo J (2022) Mental health prediction using machine learning: taxonomy, applications, and challenges. Appl Comput Intell Soft Comput 2022:1–19
2. Iqbal J, Asghar MZ, Ashraf MA, Yi X (2022) The impacts of emotional intelligence on students' study habits in blended learning environments: the mediating role of cognitive engagement during COVID-19. Behav Sci 12(1):14
3. Sujal BH, Neelima K, Deepanjali C, Bhuvanashree P, Duraipandian K, Rajan S, Sathiya-narayanan M (2022) Mental health analysis of employees using machine learning techniques. In: 2022 14th international conference on communication systems & networks (COMSNETS). IEEE, pp 1–6
4. Kour H, Gupta MK (2022) A hybrid deep learning approach for depression prediction from user tweets using feature-rich CNN and bi-directional LSTM. Multimed Tools Appl 81(17):23649–23685
5. Zeberga K, Attique M, Shah B, Ali F, Jembre YZ, Chung TS (2022) A novel text mining approach for mental health prediction using Bi-LSTM and BERT model: computational Intelligence and Neuroscience
6. Zhang T, Schoene AM, Ji S, Ananiadou S (2022) Natural language processing applied to mental illness detection: a narrative review. NPJ Dig Med 5(1):46
7. Abba SI, Abdulkadir RA, Sammen SS, Pham QB, Lawan AA, Esmaili P, Malik A, Al-Ansari N (2022) Integrating feature extraction approaches with hybrid emotional neural networks for water quality index modeling. Appl Soft Comput 114:108036
8. Garg A, Parashar A, Barman D, Jain S, Singhal D, Masud M, Abouhawwash M (2022) Autism spectrum disorder prediction by an explainable deep learning approach. Comput Mater Continua 71(1):1459–1471
9. Aleem S, Huda NU, Amin R, Khalid S, Alshamrani SS, Alshehri A (2022) Machine learning algorithms for depression: diagnosis, insights, and research directions. Electronics 11(7):1111
10. Sumathy B, Kumar A, Sungeetha D, Hashmi A, Saxena A, Kumar Shukla P, Nuagah SJ (2022) Machine learning technique to detect and classify mental illness on social media using lexicon-based recommender system. Comput Intell Neurosci
11. Ansari L, Ji S, Chen Q, Cambria E (2022) Ensemble hybrid learning methods for automated depression detection. IEEE Trans Comput Soc Syst 10(1):211–219
12. Banna MHA, Ghosh T, Nahian MJA, Kaiser MS, Mahmud M, Taher KA, Hossain MS, Andersson K (2023) A hybrid deep learning model to predict the impact of COVID-19 on mental health from social media big data. IEEE Access 11:77009–77022
13. Mohamed ES, Naqishbandi TA, Bukhari SAC, Rauf I, Sawrikar V, Hussain A (2023) A hybrid mental health prediction model using support vector machine, multilayer perceptron, and random forest algorithms. Healthc Anal 3:100185

14. Ahmad M, Wahid N, Hamid RA, Sadiq S, Mehmood A, Choi GS (2022) Decision level fusion using hybrid classifier for mental disease classification. Comput Mater Continua 72(3)
15. Shahid W, Farhan S (2022) The effect of self-compassion intervention on academic motivation and academic stress on mental health of students in hybrid learning. J Prof Appl Psychol 3(2):165–181
16. https://www.kaggle.com/datasets/osmi/mental-health-in-tech-survey

Supporting Smart Meter Context Management Using OWL Ontology and Hyperledger Fabric Blockchain

N. Sundareswaran, S. Sasirekha, M. Vijay, and K. Vivekrabinson

Abstract The usage of electrical and electronic appliances is on the rise in both homes and businesses. The smart energy device has various potential applications, including power measurement, power control, and data exchange between smart power plants and individual customer endpoints. However, the current smart energy meters primarily provide data on the overall electricity consumption of a home or business, without considering context or information security. To effectively manage energy, it is essential to have a knowledge interpreter and a secure information storage system, as most households and industries lack awareness of energy consumption, data privacy, and actions that can reduce demand. Hence, this study proposes a context-aware smart energy metering system and a secure information storage management system based on blockchain. Moreover, we analyzed the Sustainable Data for Energy Disaggregation (SustDataED2) dataset. Similarly, the Hyper Ledger Fabric (HLF) blockchain system functions as a storage ledger, ensuring the integrity of information and protecting it against malicious attacks.

Keywords Blockchain · Context management · Energy meter · Semantic rules · Hyperledger fabric · Cyber-attacks

1 Introduction

Electricity serves as the driving force behind the progress of any nation, and India has witnessed a threefold increase in electricity consumption since the 2000s. A larger proportion of households and industries now have access to electricity, with a wide range of electrically powered devices being used. As the global demand for electricity

N. Sundareswaran (✉) · M. Vijay · K. Vivekrabinson
Kalasalingam Academy of Research and Education, Krishnankovil, Tamilnadu, India
e-mail: vethasundares@gmail.com

S. Sasirekha
National Institute of Technical Teachers Training and Research, Chennai, India
e-mail: sasirekha@nitttrc.edu.in

© The Author(s), under exclusive license to Springer Nature Singapore Pte Ltd. 2024   29
D. Giri et al. (eds.), *Proceedings of the Tenth International Conference on Mathematics and Computing*, Lecture Notes in Networks and Systems 963,
https://doi.org/10.1007/978-981-97-2069-9_3

rises swiftly across residential, commercial, and industrial sectors, it has become crucial for utility companies to develop improved, non-intrusive, environmentally friendly methods of measuring utilities consumption to ensure accurate billing. Smart energy meters have been installed for precisely this purpose. These electronic devices record energy consumption on an hourly or more frequent basis and automatically generate corresponding bills [1]. Compared to traditional non-smart electrical meters, smart meters offer the advantage of providing a breakdown of electricity consumption by individual devices. The percentage of households with access to electricity has risen from 55% in 2010 to over 90% in 2022. On average, an electrified household consumes approximately 90 units (kWh) per month [2].

Electricity bills constitute a significant portion of both household and industrial budgets. Moreover, the cost per unit of electricity increases significantly once a certain consumption threshold is crossed. The aim of this work is to assist users in monitoring their electricity consumption, ensuring that only necessary devices are operational at any given time in both households and industries. This approach can effectively reduce the overall units consumed, resulting in a lower electricity bill. In the context of the smart grid network technology, the integration of smart meters with a data management system over the IoT communication network is a common practice [3, 4]. However, it is important to recognize that the smart meter system can potentially serve as an entry point for attackers seeking to disrupt the functions of the smart power system. While customer data and meter readings are typically encrypted using Advanced Encryption Standard (AES) and transmitted to the data center, relying solely on AES does not guarantee protection against all cyber-attacks related to smart meter data communication. One example is the side-channel attack, which exploits social engineering techniques to steal encryption keys. In such cases, attackers may employ bots to decrypt the information and send it to remote locations or manipulate the data to create confusion in communication [5, 6]. Moreover, the real-time transmission of smart meter readings and their relay to the destination network are susceptible to cyber-attacks [7].

There has been the emergence of a novel form of Denial of Service (DoS) attack called the puppet attack, observed in multiple countries. This attack method leverages resource exhaustion by flooding the system with excessive packets, leading to a disruption in service availability [8]. Figure 1 illustrates the scenario of cyber-attacks on a smart meter system. As a result, it is imperative to ensure the security of smart meter readings both during transmission and when stored in the database. This study employs Hyperledger Fabric (HLF) as a Blockchain platform to ensure secure storage and information operations. Within the HLF network, a storage ledger is maintained in the form of a key-value structure, which captures the current state or recent state of the network. Each block in the ledger contains essential information, including actual data, transaction details, and the latest state of the network, known as the world state. The storage ledger comprises two main components: the state database and the transaction log. The state database stores the actual data at the current state, while the transaction log serves as a library of key-value pairs, facilitating the execution of query operations. It is important to note that the storage distributed ledger within HLF

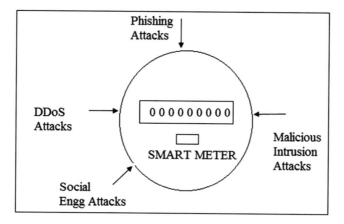

Fig. 1 Cyber attacks—smart meter system

is designed to be secure, ensuring immutability and protection against any malicious attempts to update or alter the stored information [9–11].

1.1 Motivation

Through an extensive review of existing literature, it is evident that many smart metering systems focus on managing appliance usage through statistics, automating billing processes, retrieving energy consumption status via registered mobile numbers, calculating excess electricity usage, and enhancing data storage and access systems. However, these systems often lack a knowledge interpreter, optimal control strategies, and secure storage and retrieval mechanisms. To address this issue, various inference-based approaches and blockchain platforms have been explored. The investigation reveals that semantic inference information, knowledge representation, relationship modeling, IoT system interfaces, and blockchain-based solutions for data validation, distributed ledger technology, data integrity, immutability, and traceability have significant potential. This motivates us to combine an ontology-based smart metering system with the Hyperledger Fabric (HLF) blockchain technology. In this proposed work, an ontology is designed as a context-aware system, incorporating identified devices and describing their features, instances, and attributes. This ontology serves as the knowledge base for the system. Rules and logical mechanisms are developed for the ontology, using a simple household example. The raw data collected from sensors, along with the knowledge base derived from inference, are stored and maintained within the HLF blockchain for enhanced security and reliability.

1.2 Objectives/Contributions

The primary objectives of this work are as follows:

1. Develop a context-aware smart metering system integrated with a secure information system based on blockchain technology.
2. Utilize rule-based inferences to optimize energy consumption and provide valuable suggestions to end-users.
3. Establish a secure, immutable, and traceable information storage ledger system using the Hyperledger Fabric blockchain framework.
4. Create a scalable solution that can maintain performance efficiency even as the user base expands.
5. Validate and compare the effectiveness of inferences for simple household appliances, such as light bulbs and fans.

2 Related Works

In their study, Kotecha et al. [12] introduced an automated energy metering system designed to operate without human intervention. The system incorporates an energy meter connected to a microcontroller, which plays a vital role in processing billing information. By leveraging the capabilities of the microcontroller, the system automatically calculates the billing details and transmits them to the user as a message. Tejaswini et al. [13] describe their work aimed at developing a KWH (Kilowatt hours) meter with the added functionality of alerting users through messages. When a user, using a registered mobile number, initiates a call to the GSM Modem, a message is automatically sent to that user containing information regarding the current energy consumption status and the corresponding bill. This feature ensures convenient and timely access to important energy usage and billing details for the users. Ashal et al. [14] proposed a smart energy meter capable of calculating electricity consumption for both residential and industrial buildings. If a user exceeds the allowable electricity consumption limit, they will receive a notification, and the power supply will be temporarily interrupted. Consumption details are stored on a web server. This approach ensures convenient access to consumption data while minimizing communication expenses.

Iglesias et al. [15] emphasized the growing interest in smart homes from homeowners and the research community in recent years. One of the driving factors for this interest is the potential for significant energy savings and reduced operational costs over the entire lifecycle of a building through the use of modern automation technology. Intelligent capabilities are implemented through a multi-agent system, enabling openness to the external world. This approach aims to alleviate the current problems associated with smart homes, paving the way for enhanced energy efficiency and user satisfaction. Sukhwani et al. [16] presented a performance model for Hyperledger Fabric, focusing on various aspects such as throughput, transaction processing, parallelization of peers, and identifying performance bottlenecks including latency and time-consuming validation processes. The study demonstrates that Hyperledger Fabric exhibits strong performance and scalability, making it suitable for a range of real-time applications, including Internet of Things (IoT), drones, smart meters in grid systems, and security applications. Lork et al. [17] introduced a framework for energy optimization in buildings that leverages an ontology and inference rules. Through the implementation of this framework, the researchers conducted a numerical case study and utilized experimental data obtained from a commercial building in Singapore. The results demonstrated notable advantages over traditional energy management approaches, showcasing significantly improved outcomes.

Bokhari et al. [18] conducted an investigation into blockchain-based smart meter reading validation schemes using the permissionless Ethereum Blockchain. The aim of this work was to ensure accurate meter readings, which were validated through the use of blockchain Proof of Authority (PoA). The study concluded that the proposed approach has improved the secure transmission of data and reduced the time complexity of smart metering implementation. By leveraging the capabilities of blockchain and incorporating the PoA consensus mechanism, the accuracy and integrity of smart meter readings were enhanced. Rojas et al. [19] proposed a novel cyber-security architecture based on blockchain technology to enhance data security in smart metering applications. The architecture adopts a multitier approach that incorporates cloud, edge, and fog computing to optimize the performance of data storage, access, and device interconnection in a distributed environment. The research findings indicate that the proposed architecture minimizes data tampering risks and enhances overall system performance.

Madhu et al. [20] introduced a security mechanism aimed at safeguarding smart meter readings against False Data Injection (FDI) attacks. The mechanism employs bit masking and salting operations to ensure data integrity and authentication. Asymmetric cryptographic algorithms, such as ElGamal, Secured Hash Algorithm (SHA-512), and digital signature algorithms, are utilized to enhance the security measures. Through experimental evaluations, the results clearly demonstrate that the implemented security mechanism effectively protects smart meter readings from FDI attacks. By employing techniques like bit masking, salting, and advanced cryptographic algorithms, the integrity and authenticity of the data are maintained, mitigating the risks posed by potential FDI attacks. Yan et al. [21] presented trustful data collection architecture for smart meter systems based on Hyperledger blockchain. The architecture utilizes blockchain smart contracts to automate data collection processes. The performance of the system is evaluated considering factors such as data loss, data sharing, and throughput of smart contract execution, energy efficiency, and security. The study concludes that the proposed system architecture is highly suitable for achieving traceable and reliable real-time data collection and storage from smart metering systems.

3   Proposed System Design

The combination of the proposed IoT system architecture and Blockchain Hyperledger Fabric forms a unified and robust architecture, leveraging the capabilities of both technologies. This integration is facilitated through a wireless communication interface, as illustrated in Fig. 2. Hyperledger Fabric (HLF) is acknowledged for its scalability, offering a solution to mitigate performance degradation when the ledger size or computing nodes expand. The proposed architecture incorporates a well-defined storage and security framework, which operates as follows.

In stage 1, the set consists of a group of Internet of Things (IoT) devices, led by a Head node. Each IoT device is given a unique identity by the Hyperledger Fabric (HLF) system, which automatically generates addresses. Within the architecture's perception layer, the temperature sensor (LM35), PIR sensor (HC-SR501), and current sensor (ACS 712) are all uniquely recognized by assigning them HLF addresses. Additionally, the connected devices are also assigned HLF addresses to ensure unique identification. In stage 2, the data gathered from IoT devices is directed and stored in the ledger, and access to this data is restricted to authorized participant nodes within the network through a smart contract solution. In the architecture's context and inference layers, the ontology context description of IoT data, along with its associated concepts, properties, and semantic inference information such as instances, knowledge, groups, resource descriptions, and schema, are also stored in the Hyperledger Fabric (HLF) ledger. This information is then shared exclusively with authorized participants, ensuring secure and controlled access.

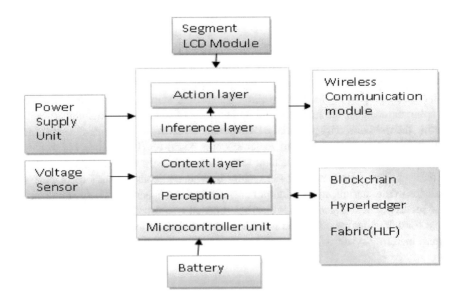

Fig. 2 System architecture

In stage 3, to ensure the security of the Hyperledger Fabric (HLF) nodes and ledger, various security controls are implemented, including authorization and authentication measures. These controls encompass data ownership and checksum verification, which contribute to data immutability. Furthermore, data pointers are stored to enable data traceability within the ledger. As an additional layer of security, asymmetric encryption/decryption algorithms are employed during data transmission, safeguarding IoT data both during transit and while at rest. Within the architecture's Action layer, the relationships among devices, sensors, people, and their objects are regarded as highly sensitive information. The aforementioned security controls implemented within HLF facilitate secure communication, interfaces, and traceability of objects throughout the architecture. This ensures that the sensitive information associated with devices, sensors, people, and their object relations remains protected and well-maintained within the system. In stage 4, the ledger is built using the robust CouchDB database, which is capable of handling large volumes of data. To facilitate data analytics, the shared ledger incorporates data analytic controls, allowing query operations through big data tools. With this capability, it becomes possible to query various types of data, including raw sensor data from LM35, HC-SR501, and ACS712 sensors, as well as information pertaining to connected devices, nodes, and historical data. The HLF-Bigdata tools provide the means to access and query these diverse data sources within the Hyperledger Fabric ecosystem.

4 Results and Discussion

4.1 SustDataED2 Dataset

The dataset consists of measurements of electrical energy consumption recorded by Portuguese householders. It provides information on household occupancy at the room level and specific appliance consumption loads. The dataset, known as SUSDataED, is available in the Single Unit Retrieval Format (SURF) file format and includes various specifications such as current and voltage at 12.8 Kilo Hertz (kHz) and 50 Hz, real and active power at 50 Hz, and plugwise measurements at 0.5 Hz stored in a SQLite database. The dataset contains power readings and event quantities for multiple appliances, including a TV, laptop, PlayStation, stove, washing machine, oven, refrigerator, freezer, coffee maker, and more [22]. Siddiqui et al. [23] utilized a Relational Database (RDB) dataset consisting of smart meter readings for their study on a knowledge-based battery drain reducer for smart meters.

4.2 Ontology Created Using Protege

The various appliances commonly found in a typical household have been identified and cataloged. To organize and represent these devices, ontology was created using the Protégé editor. In the ontology, each device is represented as a class or concept, along with their associated properties, which describe the various features and attributes of each device. These properties are often referred to as slots, roles, or properties. Additionally, restrictions on these slots can be specified to further define the relationships between concepts. When combined with individual instances of these classes, the ontology forms a knowledge base. For the ontology representation, the system utilizes the OWL (Web Ontology Language), which is a Semantic Web language designed to capture rich and complex knowledge about entities, groups of entities, and the relationships between them. While the proposed system focuses on household devices, the inferences derived from its results can be extrapolated and applied to industrial operations involving heavy electrical and electronic devices. The insights gained from this system have broader applications beyond residential settings.

4.3 Complete Ontology

The ontology includes several basic classes as shown in Fig. 3 that have been identified: House, Person, Watt, and Module. Each class serves a distinct purpose within the ontology:

1. House: This class represents residential households, encompassing all relevant
    attributes and properties associated with a house.

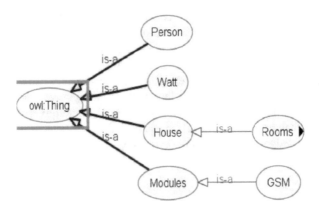

Fig. 3 Classes and subclasses

2. Person: This class represents a collection of individuals living within the house, providing a means to define and describe the residents.
3. Watt: This class signifies the unit of power consumption attributed to all electrical and electronic appliances present in the house.
4. Module: This class specifically represents the GSM module utilized for sending messages to the user. It is interfaced with the Arduino Uno microcontroller.

4.4 Ontology Rules

The ontology's rules are implemented using the owlready2 package in a Python script. These rules are written in the Semantic Web Rule Language (SWRL). SWRL is a language designed for the semantic web, allowing the expression of rules and logical statements. The rules follow an implication format, consisting of an antecedent (body) and a consequent (head). The rules are written to establish relationships and infer new knowledge based on existing information within the ontology. By leveraging SWRL, the rules enable reasoning and logical deductions to be applied to the ontology's entities and their properties, facilitating more advanced data analysis and decision-making processes [24].

4.5 Blockchain Results

As shown in Fig. 4, the Hyperledger blockchain employs CouchDB as a storage mechanism to securely store data and inferences.

In Fig. 5, the implementation of Transport Layer Security (TLS) is evident. TLS ensures secure communication by encrypting data transmitted between the client and server. Additionally, a Certificate Signing Request (CSR) is utilized to establish a secure connection between the Secure Socket Layer (SSL) and the Certificate Authority (CA). During the verification process of Hyperledger Fabric (HLF), the necessary public key and private key are generated and issued.

The above comparison and results produced by existing system Reda et al. [25] illustrated that the proposed system exhibited best performance in terms of processing time for 6 datasets as shown in Fig. 6.

```
558aba674f21    couchdb:3.1.1                       "tini -- /docker-ent…"   19 seconds ago
.0:5984->5984/tcp, :::5984->5984/tcp
24134c1669f8    couchdb:3.1.1                       "tini -- /docker-ent…"   19 seconds ago
.0:7984->5984/tcp, :::7984->5984/tcp
4871670a90a7    hyperledger/fabric-orderer:latest   "orderer"                19 seconds ago
::7050->7050/tcp, 0.0.0.0:7053->7053/tcp, :::7053->7053/tcp, 0.0.0.0:17050->17050/tcp, :::
sundareswaran@sundareswaran-Vostro-2520:~/fabric-samples/test-network$
```

Fig. 4 Couchdb creation in HLF

```
2021/10/20 11:19:25 [INFO] TLS Enabled
2021/10/20 11:19:25 [INFO] generating key: &[A:ecdsa S:256]
2021/10/20 11:19:25 [INFO] encoded CSR
2021/10/20 11:19:26 [INFO] Stored client certificate at /home/sundareswaran/fabric-samples/test-network/organizations/ordererOrganizations/exa
mple.com/users/Admin@example.com/msp/signcerts/cert.pem
2021/10/20 11:19:26 [INFO] Stored root CA certificate at /home/sundareswaran/fabric-samples/test-network/organizations/ordererOrganizations/ex
ample.com/users/Admin@example.com/msp/cacerts/localhost-9054-ca-orderer.pem
2021/10/20 11:19:26 [INFO] Stored Issuer public key at /home/sundareswaran/fabric-samples/test-network/organizations/ordererOrganizations/exam
ple.com/users/Admin@example.com/msp/IssuerPublicKey
2021/10/20 11:19:26 [INFO] Stored Issuer revocation public key at /home/sundareswaran/fabric-samples/test-network/organizations/ordererOrganiz
ations/example.com/users/Admin@example.com/msp/IssuerRevocationPublicKey
Generating CCP files for Org1 and Org2
```

Fig. 5 Secured communication in HLF

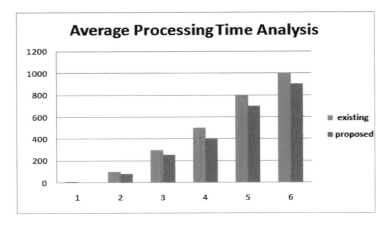

Fig. 6 Comparison of average processing time

5 Conclusion

The proposed work introduces a context-aware system designed to monitor and optimize the power usage of various household appliances. It utilizes affordable sensors that can be deployed to both residential and industrial settings. Unlike generic smart meter systems, this rule-based system can identify devices that consume excessive electricity and provide recommendations on how to monitor and optimize their usage. Users receive daily reports containing findings and personalized recommendations. To simplify the testing phase, the system was implemented using a light bulb and a small DC fan, but it can be extended to measure the electricity consumption of any AC/DC equipment in the home. Additionally, certain appliances can be automated to temporarily shut down and notify the user when consumption exceeds a predetermined threshold. Instead of daily alerts, monthly alerts with consolidated consumption information and recommendations can be delivered. To ensure the security and integrity of inference and data storage operations, the proposed system leverages the Hyperledger Fabric (HLF). HLF ensures that actual data, transaction data, and the latest state are securely stored in a distributed ledger. This ledger is resistant to malicious tampering or unauthorized modifications, providing a secure and immutable record of the data. By utilizing HLF, the proposed system guarantees the integrity and confidentiality of the collected information and ensures that users can trust the recommendations and insights provided.

References

1. Mani V, Abhilasha G, Lavanya S, Sankaranarayanan S (2017) IoT based smart energy management system. Int J Appl Eng Res 12:55–62
2. Venkatraman A, Thatte AA, Xie L (2021) A smart meter data-driven distribution utility rate model for networks with prosumers. Util Policy 70:43–57
3. Gaggero GB, Marchese M, Moheddine A, Patrone FA (2021) Possible smart metering system evolution for rural and remote areas employing unmanned aerial vehicles and internet of things in smart grids. Sensors 2:16–21
4. Chakravarthi PK, Yuvaraj D, Venkataramanan V (2022) IoT–based smart energy meter for smart grids. In: 6th international conference on devices, circuits and systems (ICDCS). India, pp 360–363
5. Rajasekaran AS, Azees M, Al-Turjman F (2022) A comprehensive survey on security issues in vehicle-to-grid networks. J Control Decis 8:1–10
6. Gui Y, Siddiqui AS, Tamore SM, Saqib F (2019) Security vulnerabilities of smart meters in smart grid. In: 45th annual conference of the IEEE industrial electronic society. Portugal, pp 3018–3023
7. Kumar P, Lin Y, Bai G, Paverd A, Dong JS et al (2019) Smart grid metering networks: a survey on security, privacy and open research issues. IEEE Commun Surv Tutor 21:2886–2927
8. Yi P, Zhu T, Zhang Q, Wu Y, Pan L (2016) Puppet attack: a denial of service attack in advanced metering infrastructure network. J Netw Comput Appl 59:325–332
9. Zhao Z (2022) Comparison of hyperledger fabric and ethereum blockchain. IEEE Asia-Pacific conference on image processing, electronics and computers (IPEC). Dalian, China, pp 584–587
10. Sammeta N, Parthiban L (2022) Hyperledger blockchain enabled secure medical record management with deep learning-based diagnosis model. Complex Intell Syst 8:625–640
11. Ravi D, Ramachandran S, Vignesh R, Falmari VR, Brindha M (2022) Privacy preserving transparent supply chain management through hyperledger fabric. Blockchain Res Appl 3:108–123
12. Kotecha V, Jadhav S, Bhisikar S, Jangda R, Kanade DM (2018) GSM technology based smart energy meter. Int J Sci Eng 3:159–162
13. Tejaswini S, Powale VB, Shende YP et al (2018) Hall effect sensor based digital smart three phase energy meter. Int Res J Eng Technol 5:1948–1952
14. Ashal R, Aruna R, Divya J, Balasaranya K (2018) Smart energy meter for advanced metering and billing alert framework. Int J Eng Sci Comput 8:541–544
15. Iglesias F, Vazquez KW (2011) Thinkhome energy efficiency in future smart homes. EURASIP J Embed Syst 11:1–18
16. Sukhwani H, Wang N, Trivedi KS, Rindos A (2018) Performance modeling of hyperledger fabric (permissioned blockchain network). In: 17th International symposium on network computing and applications (NCA 2018). Cambridge, USA, pp 1–8
17. Lork C, Choudhary V, Hassan NU, Tushar W, Yuen C et al (2019) An ontology based framework for building energy management with IoT. Electronics 485:10–23
18. Bokari ST, Aftab T, Nadir I, Bakhshi T (2019) Exploring blockchain secured data validation in smart meter readings. In: 22nd International multi topic conference (INMIC). Islamabad, Pakistan, pp 1–7

19. Rojas JCO, Archundia ER, Gnecchi JAG, Jacobo JC, Murueta JWG (2020) A novel multi-tier blockchain architecture to protect data in smart metering systems. IEEE Trans Eng Manag 67:1271–1284
20. Madhu A, Prajeesha P (2021) Prevention of FDI attacks in smart meter by providing multi-layer authentication using ElGamal and SHA. In: 5th International conference on computing methodologies and communications (ICCMC). Erode, India, pp 246–251
21. Yan L, Angang Z, Hauiying S, Lingda S, Guang S et al (2021) Blockchain based reliable collection mechanism for smart meter quality data. In: International conference on artificial intelligence and security (ICAIS 2021), pp 476–487
22. Pereira L, Costa D, Ribeiro M (2022) A residential labeled dataset for smart meter data analytics. Sci Data 134:1–11
23. Siddiqui IF, Jin lee SU, Abbas A (2020) A novel knowledge-based battery drain reducer for smart meters. Intell Autom Soft Comput 6:107–119
24. Park H, Shin S (2023) A proposal for basic formal ontology for knowledge management in building information modeling domain. Appl Sci 13:48–59
25. Reda R, Carbonaro A, Boer V, Siebes R, Weerdt R et al (2022) Supporting smart home scenarios using OWL and SWRL rules. Sensors 22:4131–4150

Modeling Vegetation Dynamics: Insights from Distributed Lag Model and Spatial Interpolation of Satellite Derived Environmental Data

Janani Selvaraj and Prashanthi Devi Marimuthu

Abstract The study proposes a method for modelling vegetation dynamics by combining time series analysis of the Normalised Difference Vegetation Index (NDVI) with spatial interpolation of environmental data. The goal is to provide a comprehensive understanding of how vegetation responds to changing environmental conditions by taking both temporal and spatial aspects into account. To investigate the temporal patterns of NDVI, advanced time series analysis techniques are used in the temporal domain. Distributed Lag Models, in particular, are utilised for modelling to discover the complex interactions between satellite derived NDVI and environmental factors such as Land Surface Temperature and precipitation. This method aids in assessing the delayed impacts of environmental influences on vegetation providing information on both short-term and long-term responses. Simultaneously, spatial interpolation methods are used in the spatial domain to build continuous maps of environmental variables across the study area. These spatial surfaces provide useful information on the geographic variation of environmental conditions. These findings have implications for ecosystem management, assessing climate change, and planning land use, providing a solid platform for informed decision-making in complex ecological systems.

Keywords Spatial interpolation · NDVI · Distributed lag model · Climate change

J. Selvaraj (✉)
Department of Data Science, Bishop Heber College, Tiruchirappalli, Tamil Nadu, India
e-mail: janani6490.selva@gmail.com

P. D. Marimuthu
Department of Environmental Science and Management, School of Environmental Sciences, Bharathidasan University, Tiruchirappalli, Tamil Nadu, India

© The Author(s), under exclusive license to Springer Nature Singapore Pte Ltd. 2024
D. Giri et al. (eds.), *Proceedings of the Tenth International Conference on Mathematics and Computing*, Lecture Notes in Networks and Systems 963,
https://doi.org/10.1007/978-981-97-2069-9_4

41

1   Introduction

Understanding the dynamics of the Earth's ecosystems is critical for informed decision-making in a variety of sectors, including ecosystem management and climate change assessment. To acquire a thorough knowledge of these intricate dynamics, researchers have increasingly recognised the need of incorporating both temporal and geographical aspects into their investigations. Vegetation is the essential foundation of ecosystem services and functions, and its dynamics serve as a sensitive indicator of ecosystem conditions, as they are strongly linked to climate change, hydrological conditions, and human activities. Examining the relationship between vegetation and climate against the context of global warming has enormous theoretical study value [1]. As the primary driver of vegetation, temperature and precipitation, has a close relationship with hydrothermal conditions. It has been a prominent subject for studying the interaction between climate and vegetation change [2].

Climate change is well acknowledged to alter vegetation development and dynamics and vegetation response to climate is frequently complex and temporal [3]. Climate, in particular, has an uneven effect on vegetation, with or without time-lag [4] and time-accumulation effects [5]. A study conducted by Ding et al. [6] was to identify the worldwide spatiotemporal responses of plants to climate. It was claimed that climatic conditions have time-lag and accumulation effects on global vegetation growth, as well as combined effects, and that these temporal effects differ between vegetation kinds, regions, and plant development phases [6].

The interaction of vegetation and climatic elements (its driving variables) has a non-linear behaviour that is influenced by time lag and time accumulation. Temperature's time-lag and time-accumulation impacts, as well as precipitation's time-accumulation impact, all have an impact on vegetation growth. The application of the Granger Causality (GC) Test and GC-based Vector Auto-Regressive Neural Network (VARNN) Model test to the climate-vegetation response mechanism demonstrates that the 0–2 month optimum time lag effect is prominent in the Jharkhand area [7].

A time series based study in China's Mu Us Sandy Land used NDVI data to analyse how climate change and human activities influenced vegetation. It discovered that variations in NDVI values were related to changes in climate parameters such as temperature and precipitation, whereas human activities such as urbanisation and land use changes influenced plant cover [8]. A study conducted in the Haryana State, India, spanning 2000 to 2022, examined vegetation dynamics using a number of data sources. It indicated elevation-related regional changes, with higher elevations enjoying more rainfall and vegetation and lower elevations experiencing increasing temperatures and less vegetation as a result of human activity. Rainfall had a substantial positive relationship with vegetation, however temperature had a negative relationship. Importantly, the study found that climate change had a greater impact at higher elevations, while human activity was the primary cause of vegetation loss in lower places [9].

Lagged regression is important for modelling vegetation dynamics because it takes into consideration temporal dependencies, response times, seasonality, and ecosystem resilience, thereby improving our understanding of how plant responds to environmental changes. This method is very useful in ecological studies and ecosystem management, as it allows researchers to make more informed decisions and predictions.

Distributed Lag Models (DLM's) are a type of regression model in which the lags of explanatory time series are used as independent variables. They provide a versatile method for incorporating independent series into dynamic regression models. DLM's are dynamic models in the sense that they include both dependent and independent series past values. Although DLMs are most commonly employed in econometric analyses, because they enable regression-like analysis for time series data, they are also widely used in energy, marketing, agricultural, epidemiology, and environmental research [10]. The Autoregressive Distributed Lag Models (ARDLM's) have been used widely in econometrics for the investigating the association of government expenditure and financial development in environmental degradation [11].

For the present study Distributed Lag Models and spatial interpolation techniques have used to understand the spatio-temporal affects of Land Surface Temperature and Precipitation on the vegetation indices.

The paper is organised as follows: The data sources section (Sect. 2) delves into the collection and processing of NDVI and environmental data. The methodology section (Sect. 3) expands on the novel approach, which combines temporal analysis with ARDL models with spatial analysis utilising spatial interpolation techniques. The results part (Sect. 3) presents major findings from this detailed research, while the discussion section (Sect. 4) explores the importance and ramifications of the findings.

2 Study Area and Data Sources

2.1 Study Area

The Karaivetti Wetland region in the Ariyalur district has been chosen as the study area. It is an ecologically significant wetland in Southern India, which is recognized for its distinct and diverse habitat. This wetland is significant in the local ecology because it supports a broad range of plants and animals while also serving as a vital water source for the surrounding towns. The research area's significant geographical and environmental qualities make it an ideal location for investigating the complicated relationships between environmental variables and vegetation indices (Figs. 1 and 2).

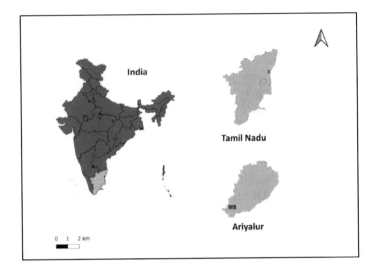

Fig. 1 Study area

Fig. 2 Karaivetti wetland region

2.2 Data

To undertake geospatial analysis, the Google Earth Engine (GEE) platform makes use of Sentinel-2 satellite imagery. It begins by filtering a set of Sentinel-2 images, selecting only those with little cloud cover and falling within a time window and research region boundaries. Following that, a mosaic is formed by calculating the median pixel reflectance values across these filtered images. For optimal visualisation, this composite mosaic is trimmed to the defined region and scaled to the 0–1 range. The Normalised Difference Vegetation Index (NDVI) [12] is calculated using the Near-Infrared (NIR) and Red bands, and it is a useful tool for monitoring vegetation health and land cover changes. The resulting picture and NDVI data can help with a variety of ecological and environmental assessments.

$$NDVI = \frac{NIR - Red}{NIR + Red}$$

The MODIS Land Surface Temperature (LST) data for the year 2017 to 2021 was collected for the study region using Google Earth Engine. The LST values are extracted, transformed, and used to create a time series values illustrating mean temperatures over the time period. Additionally, the data points are sampled for export, enabling further analysis and visualization. Similarly, the precipitation data was sourced from Climate Hazards Group InfraRed Precipitation with Stations (CHIRPS) dataset from Google Earth Engine for the time period for the study region.

3   Methodology and Results

3.1   Spatial Interpolation

The study area's NDVI (Normalized Difference Vegetation Index), Land Surface Temperature (LST), and precipitation data were obtained through satellite imagery from Google Earth Engine. These raster datasets were transformed into point datasets and subsequently downloaded as shapefiles from Google Earth Engine. These shapefiles were then imported into R Studio for further analysis. To assess the relationship between Land Surface Temperature and Rainfall with NDVI values, a simple linear model was employed. The model aimed to quantify the influence of Land Surface Temperature and Rainfall on NDVI. The resulting fitted NDVI values from this model were integrated into the shapefile dataset. In the next step, the point data containing actual and predicted NDVI values was used to generate a raster image. This interpolation process provided a spatially continuous representation of NDVI across the study area. The resulting images showcase the comparison between actual and predicted NDVI values for the years spanning from 2017 to 2021 which is presented in Figs. 3 and 4.

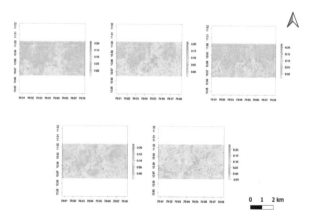

Fig. 3 Spatio-temporal variation of NDVI from 2017 to 2021

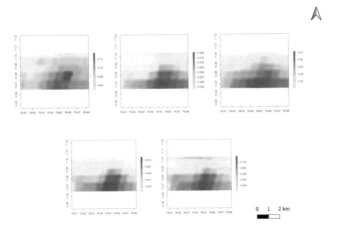

Fig. 4 Spatio-temporal variation of predicted NDVI from 2017 to 2021

3.2 *Distributed Lag Model*

The Distributed Lag Model and the Autoregressive Distributed Lag (ARDL) Model are statistical techniques that are used to examine the relationship between a dependent variable and its lagged values of explanatory factors across time.

Consider a response time series y_t and an input (or independent variable or features) time series x_t. Consider the models of the form:

$$y_t = \alpha + \sum \beta_j x_{t-j} + \varepsilon_t$$

where x_{t-j} represents the lagged values of the imput series and ε represents an i.i.d noise process. In this case the collection $\{\beta_j\}$ as a function of the lag j is referred to as the *distributed lag function*.

The Autoregressive Distributed Lag Model is represented in the form:

$$y_t = \alpha + \sum \varphi_i y_{t-j} + \sum \beta_j x_{t-j} + \varepsilon_t$$

where y_{t-j} represents the lagged dependent variable, φ_i represents the autoregressive coefficients of the lag terms.

To investigate the temporal influence of Land Surface Temperature (LST) and Rainfall on NDVI, lagged models were applied. The dLagM R package was used for performing the Distributed Lag Models [13].

The figures (Fig. 5) illustrate the temporal profiles of NDVI, LST, and Precipitation within the study area. The temporal data for these three critical variables was sourced directly from Google Earth Engine. While addressing the issue of missing data, imputation techniques were employed to fill in the gaps, ensuring a comprehensive and reliable dataset for the analysis.

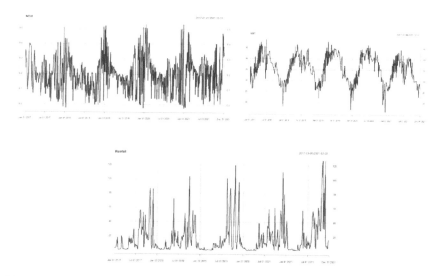

Fig. 5 Temporal trend of NDVI, land surface temperature and precipitation

The graphs representing the dynamics of NDVI, LST, and Rainfall throughout time demonstrate an interesting temporal pattern. This common temporal pattern indicates a close relationship between these environmental elements and vegetation dynamics. While the separate datasets may vary due to short-term impacts, the overall congruence of their trends during this time period highlights the considerable influence of LST and Rainfall on NDVI. This coordination provides emphasis on the seasonal and yearly cycles that influence vegetation health and highlights the potential for these patterns to feed predictive models and guide ecosystem management decisions.

Auto-regressive components are used in ARDL models, which improves their ability to capture complicated temporal connections. They provide statistical significance checks for coefficients, which ensures strong findings and builds trust in conclusions. ARDL models are more interpretable, revealing lagged impacts and temporal patterns. Their model structure is relatively explicit, making it easier to evaluate assumptions. These improvements make ARDL models a versatile and robust choice for simulating complex ecological dynamics, with the potential for enhanced predictions and insights.

The Distributed Lag Model and the ARDL Model were first applied to observe individual temporal effect of LST and Precipitation on NDVI. The summary of the DLM for the effect of LST on NDVI is presente din Table 1. It could be observed that the current LST coefficient has a statistically significant positive influence on NDVI. Notably, Lag-5 LST has a strong and negative relationship with NDVI, although data does not support the impact of previous delayed LST values. These findings shed light on the temporal dynamics of LST-NDVI interactions in the studied area.

Table 1 Summary of the DL model for the effect of LST on NDVI

	Estimate	Std. error	t value	p-value
Intercept	3.694e-01	1.418e-02	26.055	<2e-16 ***
LST	3.452e-03	1.338e-03	2.580	0.00995 **
Lag-1 LST	1.189e-04	1.711e-03	0.070	0.94459
Lag-2 LST	−1.023e-04	1.716e-03	−0.060	0.95247
Lag-3 LST	−2.885e-05	1.712e-03	−0.017	0.98655
Lag-4 LST	−1.298e-03	1.708e-03	−0.760	0.44738
Lag-5 LST	−6.970e-03	1.333e-03	−5.229	1.9e-07 ***

Table 2 summarises the results of a Distributed Lag Model (DL model) investigation into the impact of precipitation on the Normalised Difference Vegetation Index (NDVI). The significance of crucial coefficients is highlighted in the table. The current Precipitation coefficient is not statistically significant, indicating that current Precipitation has no effect on NDVI. Only Lag-5 Rainfall has statistical significance with a positive coefficient, indicating a significant positive association between Lag-5 Rainfall and NDVI. The other lag Rainfall terms are not statistically significant. These findings provide useful insights into the study area's complex interaction between precipitation and NDVI changes, with Lag-5 Rainfall having a crucial role.

Table 3 summarises the results of an AutoRegressive Distributed Lag (ARDL) model used to investigate the effect of Land Surface Temperature (LST) on the Normalised Difference Vegetation Index (NDVI). The table emphasises the importance of critical coefficients. Current LST (0.0017822) has a statistically significant coefficient (p-value: 0.009101), indicating a positive link between current LST and NDVI. Furthermore, Lag-1 LST displays a significant negative connection with NDVI (p-value: 0.000271), showing that the LST from the prior time period has a negative impact on current NDVI. Notably, Lag-1 NDVI has an exceedingly high positive coefficient (0.8573419), a t-value of 71.913, and a low p-value, showing a substantial positive temporal autocorrelation in NDVI. These findings provide important insights into the temporal dynamics of LST-NDVI interactions, emphasising

Table 2 Summary of the DL model for the effect of precipitation on NDVI

	Estimate	Std. error	t value	p-value
Intercept	0.2234767	0.0029677	75.303	<2e-16 ***
LST	−0.0006973	0.0007502	−0.929	0.353
Lag-1 rainfall	0.0004706	0.0015018	0.313	0.754
Lag-2 rainfall	−0.0005532	0.0015991	−0.346	0.729
Lag-3 rainfall	−0.0005825	0.0015991	−0.364	0.716
Lag-4 rainfall	−0.0018337	0.0015018	−1.221	0.222
Lag-5 rainfall	0.0016501	0.0007501	2.200	0.028 *

Table 3 Summary of the ARDL model for the effect of LST on NDVI

	Estimate	Std. error	t value	p-value
Intercept	0.0535724	0.0082399	6.502	1.02e-10 ***
LST	0.0017822	0.0006826	2.611	0.009101 **
Lag-1 LST	−0.0024936	0.0006835	−3.648	0.000271 ***
Lag-1 NDVI	0.8573419	0.0119219	71.913	<2e-16 ***

Table 4 Summary of the ARDL model for the effect of precipitation on NDVI

	Estimate	Std. error	t value	p-value
Intercept	0.0354803	0.0031186	11.377	<2e-16 ***
LST	−0.0006474	0.0002529	−2.560	0.0105 *
Lag-1 rainfall	0.0003295	0.0002549	1.292	0.1964
Lag-1 NDVI	0.8479540	0.0122989	68.945	<2e-16 ***

the significance of both current and lagged LST values, as well as temporal NDVI autocorrelation, in influencing NDVI.

Table 4 summarizes the results of the ARDL model investigating the influence of Precipitation on NDVI. Current LST has a negative effect on NDVI, supported by a significant coefficient. However, the impact of Lag-1 precipitation is not statistically supported. Remarkably, Lag-1 NDVI shows strong positive temporal autocorrelation. These findings shed light on the temporal dynamics of LST-NDVI interactions, highlighting the importance of current precipitation and NDVI's temporal patterns in vegetation dynamics.

Table 5 summarises the results of the ARDL model, which examined the combined influence of LST and Precipitation on the NDVI. Current LST has a favourable effect on NDVI, which is substantiated by a substantial coefficient. However, the impacts of Lag-1 and Lag-2 LST are not statistically significant. Rainfall has no effect on NDVI on its own. Lag-1 and Lag-2 Rainfall have no statistical significance. Surprisingly, Lag-1 NDVI has a substantial and highly significant positive autocorrelation.

4 Discussion and Conclusion

A detailed research of vegetation dynamics was conducted in the Karaivetti wetland, Ariyalur district, using a variety of advanced analytical tools in this study. Using satellite-derived environmental data, specifically Land Surface Temperature (LST) and Precipitation, in conjunction with the Normalised Difference Vegetation Index (NDVI), useful insights into the intricate interaction between these components were achieved. The process spatial interpolation improves understanding of geographic differences in environmental circumstances, which is necessary for understanding

Table 5 Summary of the ARDL model for the combined effect of LST and precipitation on NDVI

	Estimate	Std. error	t value	p-value
Intercept	7.551e-02	9.168e-03	8.236	3.36e-16 ***
LST	1.580e-03	6.837e-04	2.310	0.0210 *
Lag-1 LST	−1.002e-03	8.800e-04	−1.139	0.2551
Lag-2 LST	−1.592e-03	6.828e-04	−2.332	0.0198 *
Rainfall	−7.944e-05	3.814e-04	−0.208	0.8350
Lag-1 rainfall	−8.444e-04	7.008e-04	−1.205	0.2284
Lag-2 rainfall	5.840e-04	3.845e-04	1.519	0.1290
Lag-1 NDVI	8.305e-01	1.286e-02	64.583	<2e-16 ***

the spatial dynamics of vegetation responses to changing environmental factors. The interpolated maps form the foundation for researchers to investigate the spatial correlations between environmental variables and NDVI, allowing them to find localised patterns and trends.

The Distributed Lag Model (DLM) analysis offer critical insights that current LST positively effects NDVI, but Lag-5 LST has a considerable negative effect, underscoring the importance of both current and lagged LST values in understanding NDVI. The impact of precipitation is more complicated, with present precipitation lacking statistical significance and Lag-5 Rainfall favourably increasing NDVI. The AutoRegressive Distributed Lag (ARDL) models emphasize the importance of both current and lagged LST values in determining NDVI. Lag-1 NDVI has significant positive temporal autocorrelation. The immediate impact of precipitation is less obvious, with LST and NDVI temporal patterns playing a larger role. Table 5 reveals that current LST has a positive impact on NDVI, while Lag-1 and Lag-2 LST have little statistical significance, and Rainfall alone has no significant affect, with Lag-1 NDVI demonstrating considerable positive autocorrelation.

In conclusion, this research provides useful information about the temporal dynamics of Land Surface Temperature (LST), Precipitation, and the Normalised Difference Vegetation Index (NDVI) in the study area. The findings highlight the significance of evaluating both present and lagged LST values for interpreting NDVI trends, as well as the complex link between Precipitation and NDVI, particularly the role of lagged rainfall data. These discoveries are critical for ecosystem management, land use planning, and climate change assessment, laying the groundwork for educated decision-making in the region's complex socio-ecological systems.

Using advanced methods like as machine learning and deep learning can help us better understand these processes, potentially allowing for more accurate predictions and a deeper examination of complicated connections. Future research might also delve into the mechanisms underlying these relationships, investigate the impact of additional environmental factors, and extend the analysis over longer time scales to find long-term patterns.

References

1. Gao J, Jiao K, Wu S (2019) Investigating the spatially heterogeneous relationships between climate factors and NDVI in China during 1982 to 2013. J Geogr Sci 29:1597–1609. https://doi.org/10.1007/s11442-019-1682-2
2. Chen T, de Jeu RAM, Liu YY, van der Werf GR, Dolman AJ (2014) Using satellite based soil moisture to quantify the water driven variability in NDVI: a case study over mainland Australia. Remote Sens Environ 140:330–338 (2014). ISSN 0034–4257. https://doi.org/10.1016/j.rse.2013.08.022
3. Anderegg WRL, Schwalm C, Biondi F, Camarero JJ, Koch GW, Litvak ME, Ogle K, Shaw JD, Shevliakova E, Williams AP (2015) Pervasive drought legacies in forest ecosystems and their implications for carbon cycle models. Science 349:528–532. https://doi.org/10.1126/science.aab1833
4. Wen Y, Liu X, Pei F, Li X, Du G (2018) Non–uniform time–lag effects of terrestrial vegetation responses to asymmetric warming. Agric For Meteorol 252:130–143. https://doi.org/10.1016/j.agrformet.2018.01.016
5. Zhang H, Liu S, Regnier P, Yuan W (2018) New insights on plant phenological response to temperature revealed from long–term widespread observations in China. Glob Change Biol 24:2066–2078. https://doi.org/10.1111/gcb.14002
6. Ding Y, Li Z, Peng S (2020) Global analysis of time-lag and -accumulation effects of climate on vegetation growth. Int J Appl Earth Observ Geoinf 92:102179. ISSN 1569-8432.https://doi.org/10.1016/j.jag.2020.102179
7. Kumar V, Bharti B, Singh HP, Topno AR (2023) Assessing the interrelation between NDVI and climate dependent variables by using granger causality test and vector auto-regressive neural network model. Phys Chem Earth Parts A/B/C 131:103428. ISSN 1474-7065.https://doi.org/10.1016/j.pce.2023.103428
8. Gao W, Zheng C, Liu X, Lu Y, Chen Y, Wei Y, Ma Y (2022) NDVI-based vegetation dynamics and their responses to climate change and human activities from 1982 to 2020: a case study in the Mu Us Sandy Land, China. Ecol Indic 137:108745. ISSN 1470–160X. https://doi.org/10.1016/j.ecolind.2022.108745
9. Banerjee A, Kang S, Meadows ME, Xia Z, Sengupta D, Kumar V (2023) Quantifying climate variability and regional anthropogenic influence on vegetation dynamics in northwest India. Environ Res 234:116541. ISSN 0013–9351. https://doi.org/10.1016/j.envres.2023.116541
10. Huffaker R, Fearne A (2019) Reconstructing systematic persistent impacts of promotional marketing with empirical nonlinear dynamics. PLoS One 14(9):e0221167. https://doi.org/10.1371/journal.pone.0221167. PMID: 31532779; PMCID: PMC6750578
11. Guo P, ul Haq I, Pan G, Khan A et al (2019) Do government expenditure and financial development impede environmental degradation in Venezuela? PLOS ONE 14(1):e0210255. https://doi.org/10.1371/journal.pone.0210255. PMID: 30629649
12. Vermote E, Justice C, Claverie M, Franch B (2016) Preliminary analysis of the performance of the Landsat 8/OLI land surface reflectance product. Remote Sens Environ 185:46–56
13. Demirhan H (2020) DLagM: an R package for distributed lag models and ARDL bounds testing. PLoS ONE 15(2):e0228812. https://doi.org/10.1371/journal.pone.0228812

Effective Data Transmission in NDN-Assisted Edge-Cloud Computing Model

Po-An Shih, Cheng-Che Wu, Chia-Hsin Huang, and Arijit Karati

Abstract The proliferation of data generated by Internet of Things (IoT) devices has prompted the pursuit of streamlined data retrieval as a fundamental objective. Although Edge computing performs computations locally while significantly reducing cloud overhead, it lacks location anonymity while integrating multiple clouds. Named Data Networking (NDN) as a novel Internet architecture provides location anonymity and enhances the efficacy of data exchange through caching. In this paper, we develop an efficient data retrieval system that protects file location privacy across multiple-cloud platforms by leveraging NDN and edge computing. In our Edge-NDN architecture, consumers perform data queries on the local network without connecting to the cloud. Initially, consumers search for edge-cacheable data; if content cannot be located, it is retrieved via an external connection to the cloud. We estimate the performance of the new work using NDN Forwarding Daemon (NFD), ndn-cxx, and jNDN tools. The empirical findings indicate that the proposed framework for facilitating anonymous data communication outperforms the conventional cloud-centric approach.

Keywords Named data networking · Edge computing · Cloud computing · Internet of Things · Decentralized security

This work was supported in part by the National Science and Technology Council (NSTC) under grants NSTC 111-2222-E-110-008 and NSTC 112-2221-E-110-027, and assisted by the CANSEC-Lab@NSYSU in Taiwan.

P.-A. Shih · C.-C. Wu · C.-H. Huang · A. Karati (✉)
National Sun Yat-sen University, Kaohsiung 80424, Taiwan
e-mail: arijit.karati@mail.cse.nsysu.edu.tw

53

1 Introduction

Due to the accelerated growth of the Internet and consumer electronics, including mobile phones and personal computers, fine-grained access control over data has become an essential aspect of contemporary life [1]. As shown in Fig. 1, internet users in developed, emerging, and global nations are proliferating annually [2] due to cloud storage [3], communication applications, live streaming, and search engines. Consequently, a vast quantity of information is exchanged on the Internet. Many network system architects and data scientists are now considering the most effective way to obtain this data and improve user experience.

The rise of IoT, Industry 5.0, and smart devices has increased data storage and processing demands for organizations and consumers [4]. To conserve device storage capacity, many users are accustomed to storing data on cloud-based hard drives. While Google, Apple, and Amazon have created cloud storage systems [5], these systems are individually susceptible and partially trustworthy. In most cloud services, customers access data through provider servers. Users must also reconnect to the service provider and access data each time. The user must reconnect and redownload cloud data even if the same file is needed. This approach would limit file transfer efficiency due to bandwidth allotted by telecom, cloud storage, etc., providers, resulting in unstable or poor transmission efficiency. Named data networking (NDN) is a recent development that may reduce data transmission latency [6]. In contrast to standard IP networks, NDN prioritizes data packets above the host server origin, allowing the node cache to fulfill user requests. A further technique for improving data transmission efficiency is edge computing architecture [7]. Edge computing is a distributed computing architecture that analyzes, computes, and makes choices on data on network terminal-proximal nodes. With edge computing, data access, and service operations do not always need to be performed in the cloud [8]; instead, data can be managed locally, lowering the transmission path of data packets and enhancing data access speed and overall transmission efficiency. Consider a scenario in which YouTube's web server receives multiple requests from three users under the traditional network design. It proceeds with three distinct responses for consumers who

Fig. 1 Internet users

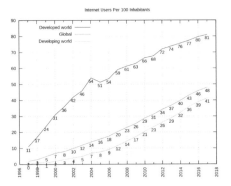

request the same video. Using Edge-NDN, when an NDN node accesses the same video, it may use caching to respond promptly to subsequent users requesting the same data. Consequently, the cloud server's duty could be reduced, and transmission efficiency could be further improved [9].

1.1 Our Contributions

We design an Edge-NDN model, integrating edge computing and caching at NDN nodes, to reduce cloud overhead and improve the data transmission experience. Specific cryptographic algorithms can also be deployed at the network's edge to filter out consumer access. Before applying a security solution, we extend this mode to support multiple-cloud platforms integration. In this model, each data producer can freely choose their cloud, and a consumer can download cloud data without knowing the exact file location and cloud information. In the proposed model, the NDN nodes are equipped with a cache. All data requested through the node will be stored at the edge. If the identical data request is made again, an edge node directly retrieves it from the cache and sends it to the consumer. Data can be viewed throughout the transfer process without connecting to the cloud service server via an external network, saving transmission time and network energy. In contrast, if the user requests data that has not been processed in the NDN, the edge node in the network will not have the cached data, necessitating a connection to the cloud storage server via the external network for retrieval and access. Based on the performance analysis, the proposed model facilitates faster data access and enhances the overall user experience.

1.2 Road Map

The rest of the manuscript is organized as follows: Sect. 2 illustrates in detail the technical prerequisites. Section 3 describes our research methods and the proposed model, whereas Sect. 4 mentions the performance evaluation. Finally, Sect. 5 concludes this work with some remarks.

2 Essential Preliminaries: Concept and Definitions

We discuss NDN functional foci on node operation and packet exchange.

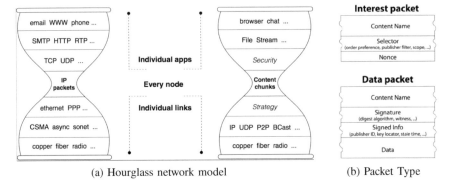

(a) Hourglass network model   (b) Packet Type

Fig. 2 Named data networking

2.1 Named Data Networking (NDN)

Figure 2a depicts the current Internet hourglass architecture [10], with the left side illustrating the current Internet Hourglass Architecture based on the Internet Protocol (IP) at the Network Layer. Currently, IP addresses play a crucial role in transmitting messages and packets, and the primary function of NDN is to convert the slender waist of an IP address into an NDN packet, as in Fig. 2b, by substituting the digital address to increase the speed of packet transmission. Table 1 compares current Internet architecture with NDN-based architecture, including protocols, operational principles, packet forwarding, and privacy.

2.1.1 Operation Mechanism of NDN Node

The NDN packet naming method adheres to the hierarchical naming convention used for URLs [11]. The names of the designating data network packets are separated by forward slashes and organized hierarchically.

Figure 2b depicts two categories of packets transmitted in NDN [12]: Interest Packets and Data Packets. In the NDN setting, a consumer sends the request to the sender via an interest packet. The packet contains the name of the data that the consumer desires. According to the research [13], each NDN node is composed of three parts, which are

1. Content Store (CS) for storing cached data.
2. Pending Interest Table (PIT) for recording interest packets
3. Forwarding Information Base (FIB) for storing and forwarding policies [14].

Figure 3a depicts the internal flow chart of the node receiving the interest packet [13]. After the node receives the Interest packet, the CS will determine whether the node possesses the required information. Such information will be returned to the preceding node or consumer if it exists. If no such data exists, however, it will be sent

Table 1 Comparison of current internet and NDN architectures

Functionalities	Internet architecture	NDN-based architecture
Protocol	TCP/IP protocol	NDN protocol: interest/data/name (IDN) and content store protocol (CSP)
Communication model	Host-to-host communication model based on IP address	Content-oriented communication model based on data names
Packet forwarding	Based on IP address	Based on matching data name
Caching	Typically limited	Caching data at various points in the network
Privacy	IP address reveal the identity and location of devices	Data names can be used to access data without revealing device identity

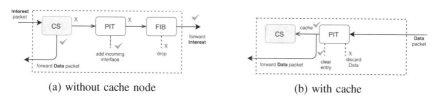

(a) without cache node   (b) with cache

Fig. 3 When the interest packet is delivered to the NDN node

to the PIT framework. The PIT architecture will check if the same Interest packet is present in the table, which involves determining if other nodes have requested the requested information. If it has already been requested, adding an interface record is unnecessary. If not, add the requestor's interface to the PIT and send it to the FIB framework. In the FIB architecture, the interfaces to which the Interest packet should be sent are determined based on the routing table and forwarding strategy.

Figure 3b depicts the internal flow chart of receiving the data packet. Suppose a data packet is passed to an NDN node. In that case, a copy of the data will be copied into the node's CS, and the interface recorded in PIT will be checked; the interface record of PIT will be deleted, and send data packets to these interfaces.

2.1.2 Packet Naming Method in NDN

The NDN architecture transmits packets based on the packet name to allow them to pass through the layered structure and be more readable. The transport layer nodes do not need the receiver's IP address to send and receive packets. In naming the package, use "/" (separator) to separate the name. For example, if an Interest packet containing *a.txt* is to be sent to other nodes, the naming method will add the data name and type and the target node to which the packet goes, such as /ndn/nsysu/word/*a.txt*.

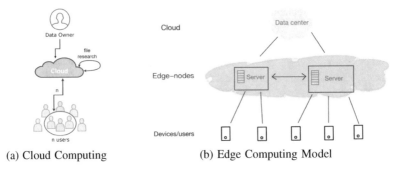

(a) Cloud Computing   (b) Edge Computing Model

Fig. 4 Architectural overview of cloud computing and edge computing

2.2 Data Authentication in NDN

According to the work [15], the identity of each data packet within an NDN packet will vary based on the data content's signature. The public key in each data packet is stored in its KeyLocator portion, which is used for signature verification of the NDN data packet. Consumers can use the trusted public key obtained in advance from the administrator to verify the data content of the data packet and the public key located in the KeyLocator to ensure the person transmitting the packet is in a trusted role. This authentication method is utilized to resolve transmission interface issues. Yu et al. [15] suggested a Trust Schema for consumers to verify data content signatures.

2.3 Architecture of Edge Computing

In the cloud computing model, as depicted in Fig. 4a, if the number of connected devices increases significantly, cloud computing will generate a large amount of data and consume a large amount of bandwidth because devices (such as mobile phones, hosts) must transmit data to the central data center for processing. Nevertheless, as depicted in Fig. 4b, if n users access the cloud concurrently, the cloud server will be inundated with n requests concurrently, resulting in a significant overhead. In this work, edge computing services trump specific cloud computing characteristics. Edge nodes can bring the computing process as close as feasible to the data source to reduce delay and bandwidth usage, thereby reducing the number of calculations and the burden on customers by directly accessing the cloud to accelerate the system's operation [16].

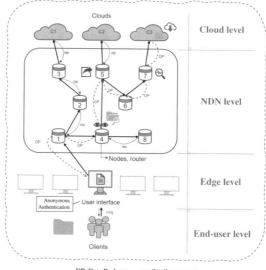

Fig. 5 The edge-cloud computing model of this study

DP: Data Package,   req: Sending request

3 Proposed Data Sharing Model

In this section, we first introduce the system model and its entities. Then we will briefly introduce the useful command for packet operations. Finally, we provide a detailed scenario of the NDN network.

3.1 System Model and Execution Steps

As shown in Fig. 5, the proposed model consists of several entities, each assigned specific roles as mentioned below.

- **Clients**: Users can send interest packets to edge nodes through the UI interface to access NDN Layer.
- **Edge nodes**: In the Edge-NDN-level, edge nodes communicate with NDN Layer and customers. In addition, there is no difference between the edge node and the node in the NDN Layer. The primary function is to provide user access and provide services to customers. The information is confirmed by the digital signature between the control devices, checks whether the PIT has records, and sometimes provides instructions for return.
- **NDN Layer**: At the NDN-level, the authenticated key is appended to the end of the name of the returned data payload as a tool. As shown in Fig. 5, the Interest packet sent by *node 1* is received by *node 2* and *node 4*. Further, add a separate

key after each interest packet (*node 2* and *node 4*) so that the returned data packet can be delivered to the user along the correct path.

- **Edge-Cloud**: When each node cannot find the required data, it will use the edge node to access the cloud for data search. Edge nodes refer to intermediate nodes 3, 5, and 7 in Fig. 5, respectively connected to cloud C1, cloud C2, and cloud C3. The FIBs in nodes 3, 5, and 7 will only have their paths to the cloud, and the edge nodes will not have other paths other than the cloud to avoid the possibility of repeated access to the cloud.

Figure 5 depicts the methodology utilized in our cloud-assisted Edge-NDN strategy. Each computer operates as a router, receiving interest packets and transmitting data packets to the front router or requester. The client functions as the requester, transmitting packets through the user interface to each router. In the guise of clouds, multiple data storage facilities exist. The data owner will first upload the data he owns to the cloud and wait for the user to use the system to search for the desired data. Assume that the connection speed for users to query the required data through various clouds directly is 1000 MB/s. The network speed required to request information from network nodes is 100 MB/s (the speed of interconnection between network nodes is also 100MB/s); according to the assumed information, if the data can be obtained by directly accessing the NDN node instead of directly accessing the cloud, the time required must be much less than directly accessing the cloud to obtain the data.

Consider a scenario where a user wishes to locate *a.txt*, and the data is stored in the C3 cloud. If the data is discovered through Edge-NDN, an interest transmission will be sent to the UI. The UI will convert the request into named data (*/nsysu/node/a.txt*) and transmit it to the connected network node (router 1). If node 1 does not have the required data, the UI will locate the next node through the routing table in the FIB and send another Interest packet. Both nodes 2 and 3 receive the same approach. If no data is found, it will search for a.txt on the C1 cloud. If no results are found, data tracing will be performed, and a NULL packet will be sent along the path to signify that an is not found on this path. After node 1 transmits the interest packet to node 4, node 4 transmits it to node 5, node 5 transmits it to node 6, and node 6 transmits it to node 7. Sends the data packet back to the user along the path and copies a.txt to the CS in each node along the path to facilitate the faster return of data when searching for data again. And if the user is looking for cloud data directly, first search for data from C1 cloud, then C2 cloud, and finally C3 cloud, the search time will be significantly longer than searching for data through Edge-NDN.

3.1.1 NDN System Setting

There are several ways of packet delivery [17]. Table 2 lists out a set of instructions for NDN tools and their respective functions.

However, in this work, we adopted *Client Control Strategy* because, like Broadcast Strategy, it must cope with NULL data packets while broadcasting interest packets. This procedure increases the load on a particular node. In the NDN architecture [18],

Table 2 Instructions of NDN tool and their functions

Commands	Functions
ndnpoke	A consumer program that sends one interest and expects one data
ndnpeek	A producer program that serves one data in response to an interest
ndnputchunks	A producer program that reads a file from the standard input and makes it available as a set of NDN data segments
ndncatchunks	A consumer program that fetches data segments of a file and writes the content of the retrieved file to the standard output

Fig. 6 Detailed flow of packet transmission and reception

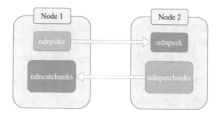

when an interest packet is conveyed to an NDN node or an edge node, the node will first store the interest packet in the PIT and then determine if the consumer has requested any data in the CS. If such data exists, the interface in the PIT will be removed, and the data packet containing the data will be sent back to the interface indicated by the PIT to complete a system search. If no such data exists, we can use *ndnpoke* in the *ndn-tools* to input the desired data name and the *ndncatchunks* on the node to continue waiting for the returned data packet. The *ndnpeek* command is used by the subsequent node to receive the interest packet, which is the previous node of the desired profile name. If a node has the required data, it will use *ndndputchunks* to convey the data payload back to the preceding node. As shown in Fig. 6, *node 1* sends an interest packet to *node 2* via *ndnpoke*, and *node 2* receives the packet via *ndnpeek*. If data is required for the interest packet, *node 2* will use the *ndnputchunks* command to send the data packet Send back to *node 1*, and *node 1* receives the file contained in the data packet through the *ndncatchunks* command.

4 Performance Evaluation and Result Discussion

We utilized Ubuntu to launch three virtual machines on a single host during our investigation. The experiment consists of a client, a server node that also functions as an NDN-level node, and an NDN-level node that connects to the cloud. In the subsequent experimental stages, users send file requests such as file1001.txt and file.pdf to the server. This experiment compares the time it takes to retrieve data from the NDN node with or without cache files to the time it takes to access the cloud directly. NDN Forwarding Daemon (NFD), ndn-cxx, and jNDN utilities are utilized to complete the project. To evaluate the efficacy of the system devised in this study, we intend to directly access cloud files via TCP/IP and utilize our NDN-based edge-cloud-sharing system. The required access time is contrasted if a single cloud file is accessed frequently.

4.1 System Execution Process

Three scenarios illustrate the process of accessing cloud files via the system.

- No nodes in the system have files, so the system must ask from the cloud.
- Edge node is liable for communicating with users and storing files.
- The node in the system farthest from the user has the file.

As shown in Figs. 7, 8, 9, 10 and 11, orange rectangles are used to separate the three functions in the system: users (left), edge nodes communicating with users (middle), and the edge node responsible for accessing the cloud (right).

4.1.1 Cloud Communication Due to Data Unavailability at the Edge

In Fig. 7, the user sends out a request for *file1001.txt* and starts accessing it one by one from the first node in the system to confirm whether the requested file exists in the node until accessing the edge node communicating with the cloud. If the edge node still does not have the file, then the edge node is responsible for accessing the cloud at this time and requests the file. As shown in Fig. 8, the edge node communicating with the cloud took 1.424 s to successfully retrieve the file from the cloud and send it back to the previous node, which also issued a demand for it. This node communicates with the user here node, so when the node receives the returned *file1001.txt* file, it

Fig. 7 The user (left) requests the system for *file1001.txt*

Fig. 8 The status and access time upon effective execution

Fig. 9 Access time when the edge node (center) obtains customer requests (left)

will also return the file to the user who requested it. Then, the total time spent by the user requesting the file, from sending the request to successfully obtaining the file, is 1547 milliseconds.

4.1.2 Edge-User Communication and Data Storage

When a user requests a file available at the nearby edge node, the file will be directly sent back to the user from this node. As shown in Fig. 9, the user sends a file request for file.pdf, and finally, he only spends 5 milliseconds to get the file.

Fig. 10 Scenario when the second node (right) has customer's desired file (left)

Fig. 11 Access time without and with data in the system

4.1.3 Farthest Edge Node from the User

In Fig. 10, the user requests *file1001.txt*, which is transmitted from the starting edge node and searched for till the farthest node (by distance). The furthest edge node accesses the cloud and has the desired file (*file1001.txt*). In contrast to scenario (1), the edge node already has the requested file, eliminating the need to request it from the cloud. Thus, *file1001.txt* will travel from this node to the user. The edge node that communicates with the user returns it along the original path. Figure 11 shows that the user obtained the file in 137 milliseconds. As shown in Fig. 11, even though the user requests the same file, it must be retrieved on the initial request due to the lack of files on each node in the system. The cloud takes 1547 milliseconds, and the second time 137 milliseconds since the first node it reaches already has a file and can be returned quickly. The former takes 11 times longer.

Fig. 12 Comparative
analysis of system access
time with and without data

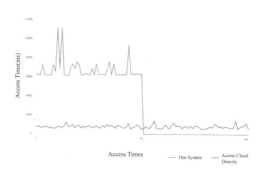

Table 3 Comparison of the results

	Our system with NO cache files	Our system with cache files	Access cloud directly
Average access time	6 s	0.006 s	0.877 s

4.2 Further Discussion

We compared the above circumstances to assess this system's functionalities. As depicted in Fig. 12, the blue line shows file access time in the proposed model. The first 50 times fulfill criterion 1, while the final 50 times meet condition 2, with 100 times of access. The orange line shows the 100-time record of direct cloud access. Downloading files directly from the cloud takes 0.877 s on average. The proposed edge-cloud system has two feasible scenarios: a) edge nodes have no file caching; fetching takes 6 s; and b) edge nodes enable file caching; accessing time is 0.006 s. One may note that direct cloud access provides 6.8 times the transmission efficiency of this method without cached data. The transmission efficiency of this technology with files is 6.8 times that of direct cloud access 146 times. Table 3 compares the average access time for our model without cache files, with cache files, and with direct cloud access. Hence, the findings show that the edge-cloud-sharing system based on the NDN described in this study increases transmission efficiency when consumers access the same data multiple times. Consequently, the proposed model has the potential to achieve decentralized security by enabling edge devices to oversee subscribers and regulate data access where clouds serve as a viable data storage alternative.

5 Conclusion and Future Work

This paper presented an edge-NDN architecture and effective data transmission combining NDN's location anonymity and caching with edge computing's local data processing and reducing multiple-cloud reliance. The architecture employs caching at the edge, allowing for rapid cached data retrieval and reducing the reliance on external connections for data access. We evaluated the performance using NDN Forwarding Daemon (NFD), ndn-cxx, and jNDN tools. Our technique outperformed the cloud-centric strategy in communication and file anonymity.

Further study can improve Edge-NDN security mechanisms, even if our approach provides location anonymity and decentralized security. Modern encryption, authentication, and access control may be used to protect data and prevent unauthorized access. Future research can combine edge nodes and several cloud storage systems to maximize edge computing and cloud resources through practical data migration approaches, ensure data consistency across edge-cloud settings, and enable dynamic resource allocation.

References

1. Karati A, Fan CI, Zhuang ES (2022) Reliable data sharing by certificateless encryption supporting keyword search against vulnerable KGC in industrial Internet of Things. IEEE Trans Indus Inform 18(6):3661–3669
2. Statista: list of countries by number of internet users 2023 (2023). https://www.statista.com/statistics/262966/number-of-internet-users-in-selected-countries/
3. Karati A, Fan CI, Zhuang ES (2021) Reliable data sharing by certificateless encryption supporting keyword search against vulnerable kgc in industrial Internet of Things. IEEE Trans Indus Inform 18(6):3661–3669
4. Al-Fuqaha A, Guizani M, Mohammadi M, Aledhari M, Ayyash M (2015) Internet of things: a survey on enabling technologies, protocols, and applications. IEEE Commun Surv Tutorials 17(4):2347–2376
5. Gubbi J, Buyya R, Marusic S, Palaniswami M (2013) Internet of things (IoT): a vision, architectural elements, and future directions. Future Gener Comput Syst 29(7):1645–1660
6. Fan CI, Karati A, Yang PS (2021) Reliable file transfer protocol with producer anonymity for named data networking. J Inf Secur Appl 59:102851
7. Shi W, Cao J, Zhang Q, Li Y, Xu L (2016) Edge computing: vision and challenges. IEEE Internet Things J 3(5):637–646. https://doi.org/10.1109/JIOT.2016.2579198
8. Satyanarayanan M (2017) The emergence of edge computing. Computer 50(1):30–39. https://doi.org/10.1109/MC.2017.9
9. Premsankar G, Di Francesco M, Taleb T (2018) Edge computing for the internet of things: a case study. IEEE Internet Things J 5(2):1275–1284. https://doi.org/10.1109/JIOT.2018.2805263
10. Named data networking: executive summary (2015). https://named-data.net/project/execsummary/
11. Zhang L, Afanasyev A, Burke J, Jacobson V, Claffy KC, Crowley P, Papadopoulos C, Wang L, Zhang B (2014) Named data networking, technical report ndn-0019. https://named-data.net/wp-content/uploads/2014/04/tr-ndn-0019-ndn.pdf
12. Lixia Z, Deborah E, Jeffrey B, Van J, James D, T., Diana KS, Beichuan Z, Gene T, kc C, Dmitri K, Dan M, Christos P, Tarek A, Lan W, Watrick C, Edmund Y (2010) Named data networking, technical report ndn-0001. https://named-data.net/techreport/TR001ndn-proj.pdf

13. Pires S, Ziviani A, Sampaio LN (2021) Contextual dimensions for cache replacement schemes in information-centric networks: a systematic review. Peer J Comput Sci 7:e418. https://doi.org/10.7717/peerj-cs.418
14. Zhang L, Afanasyev A, Burke J, Jacobson V, Claffy K, Crowley P, Papadopoulos C, Wang L, Zhang B (2014) Named data networking. SIGCOMM Comput Commun Rev 44(3):66–73. https://doi.org/10.1145/2656877.2656887
15. Yu Y, Afanasyev A, Clark D, Claffy K, Jacobson V, Zhang L (2015) Schematizing trust in named data networking. In: Proceedings of the 2nd ACM conference on information-centric networking. ACM, New York, NY, USA, pp 177-186
16. Release G (2020) What is edge computing? definition and cases explained—gigabyte global. https://www.gigabyte.com/Article/living-on-the-edge
17. Forwarding strategies—overall ndnsim 2.0. https://ndnsim.net/2.0/fw.html
18. Ndn essential tools. https://github.com/named-data/ndn-tools

Analysis of Blood Transfusion Dataset Using Data Mining Techniques

Sumathi Ganesan and G. Mahalakshmi

Abstract Data mining is the process of examining and learning from massive dataset. Here, the Blood Transfusion dataset is being processed, and classification algorithms were used to obtain the necessary knowledge. First, five classification algorithms of bagging, LogitBoost, J48, ClassificationViaRegression and Random-Forest are applied to the blood transfusion dataset. Before normalization, the values for the performance metrics of accuracy, sensitivity, specificity and error rate were found for the above classification algorithms. Then, the dataset was normalized within the values between 0.0 and 1.0 and the same performance metric values were found with the above classification algorithms. From the results of the performance metrics, RandomForest was the best classification algorithm for the blood transfusion dataset.

Keywords Blood transfusion · Classification algorithms · Normalization · RandomForest · Bagging · LogitBoost algorithm · J48 algorithm · ClassificationViaRegression

1 Introduction

The medical process of transferring blood components from one donor to another (donor to recipient) is known as a blood transfusion. Only when transfusions are required to enhance a patient's health and maybe save lives are they administered. A number of factors, including the recipient's safety, the donor and recipient's blood types' compatibility, and the type of blood products used, are taken into account before a blood transfusion is performed. Donor blood is typically easily accessible

S. Ganesan (✉)
Department of Mathematics, CEG Campus, Anna University, Chennai 600025, India
e-mail: sumisundhar@auist.net

G. Mahalakshmi
Department of Information Science and Technology, CEG Campus, Anna University, Chennai 600025, India
e-mail: mahalakshmi@auist.net

© The Author(s), under exclusive license to Springer Nature Singapore Pte Ltd. 2024
D. Giri et al. (eds.), *Proceedings of the Tenth International Conference on Mathematics and Computing*, Lecture Notes in Networks and Systems 963,
https://doi.org/10.1007/978-981-97-2069-9_6

and has a 42-day shelf life. The Blood Transfusion Service Center [1] has conducted an analysis on the blood donation service provided by individuals. The attributes of the dataset include Donor ID, Last date of donation, total donations total volume and months since first donation [1].

The tool for data mining is called Weka. The University of Waikato in New Zealand developed the Java-based machine learning suite known as the Waikato Environment for Knowledge Analysis, or Weka [2]. The Weka tool has a number of classification algorithms; the top five are chosen based on the chosen dataset. The RandomForest, Bagging, LogitBoost, Classification Via Regression, and J48 algorithms are the ones that were chosen. After processing, the algorithms' sensitivity, specificity, and accuracy were discovered. Of the five algorithms selected, RandomForest yields the best accuracy. After normalizing the dataset, the procedures are carried out once more. Normalization can be thought of as a pre-processing step, a mapping technique, or a scaling tool. The technique that provides a linear transformation on the initial data range is the Min–Max Normalization Following normalization, the outcomes are recorded, and a conclusive analysis determines if normalization is beneficial or not.

Here is an explanation of the paper's organization. Section 2 lists the several relevant blood transfusion analysis works. Section 3 provides a description of the dataset. An description of the classification algorithm techniques is given in Sect. 4. The findings and performance comparison of the Classification via Regression, Bagging, LogitBoost, J48, RandomForest, and with and without normalizing algorithms are highlighted in Sect. 5. Section 6 concludes the paper.

2 Literature Survey

This section reviews the research projects completed by different researchers who are involved in the examination of blood transfusions.

Santhanam and Sundaramin [3] employed the Classification and Regression Tree (CART) from the Blood Transfusion Service Center dataset. They showed that this method can accurately distinguish potential blood donors who have already donated (recall/sensitivity of 94%) [3]. They found the next year a comparable study comparing a Regular Voluntary Donor to a DB2K7 (Donated Blood in 2007). This study's publication by one of the original authors led to a marginal improvement in recall and precision by Sundaram in 2011. Their main accomplishment was to make the RVD model more accurate than DB2K7 [4].

An additional examination of this dataset was carried out in 2010 by Darwiche et al. [5], who assessed ANN using a radial basis function (RBF) and looked at the efficiency of Support Vector Machines. Despite the small feature space, they employ Principal Components Analysis as feature inputs as opposed to raw feature inputs in the building and assessment of these models. The SVM model performed better than the others when PCA was used as the input since it yielded the highest area under the curve (AUC) on the test set is 77.5%. The ANN model employed recency and monetary value as its attributes to achieve the best AUC of 72.5%.

Ultimately, they found that the study design used by Darwiche et al. [5] was superior to that of Santhanam and Sundaram [3] because the test set, also known as the holdout set, is used to evaluate their models, resulting in more accurate performance on subsequent observations. Furthermore, this design helps determine whether the model has over-fitted the data by comparing the statistics from the training and testing sets [5].

In order to predict future blood donation behavior, Zabihi et al. [6] collected data from the 2011 blood transfusion service center dataset (748 records/donors, 5 attributes) using fuzzy sequential pattern mining. The following features were examined in this study: (1) the number of donations made overall; (2) the time (in months) since the first donation; (3) the months since the last donation; and (4) a binary feature that indicated whether or not blood was donated in March 2007. The following are the precision/PPV features: frequency (88%), recency (72%), and time (94%) [6].

Using the Blood Transfusion Service Center dataset, Lee and Cheng's (2011) [7] study yielded the following results: AUC (72.2%), sensitivity (59.5%), accuracy (77.1%), specificity (78.1%), and yield (50 times) J48, Naïve Bayes, Naïve Bayes Tree, k-Means clustering, and Bagged ensembles of CART, NB, and NBT. This model has the best AUC when measured against other competing models [7]. Based on the aforementioned relevant works, a blood transfusion analysis is conducted by using various classification algorithms.

3 Dataset Description

We used the Blood Transfusion dataset that other researchers have made available on the UCI Machine Learning Repository in order to investigate this similar problem [6]. The source data was obtained from the Blood Transfusion Service Center's donor database in Hsin-Chu City, Taiwan [2]. Approximately 748 donors were chosen at random for the study from the donor database. One binary variable that indicates whether the donor donated blood in March 2007 (1 means blood donation; 0 means non-donation); other features that are measured are Recency months since last donation, Frequency–total number of donation, Monetary–total blood donated in c.c.), and Time–months since first donation [2] which is shown in Fig. 1.

4 Methodologies

Here, the Blood transfusion dataset is normalized into a new dataset and as a result two datasets (i.e.) normalized and unnormalized datasets are obtained. Now, various classification algorithms are applied on the datasets and based on the accuracy (%), top 5 algorithms are chosen and the metric values of both the datasets are comparatively studied. The top 5 algorithms are discussed below.

File Edit View

| Blood Transfusion.arff |

Relation: Blood Transfusion

No 1: Recency (months) 2: Frequency (times) 3: Monetary (c.c. blood) 4: Time (months) 5: whether he/she donated blood in March 2007

	Numeric	Numeric	Numeric	Numeric	Numeric
1	2.0	50.0	12500.0	98.0	1.0
2	0.0	13.0	3250.0	28.0	1.0
3	1.0	16.0	4000.0	35.0	1.0
4	2.0	20.0	5000.0	45.0	1.0
5	1.0	24.0	6000.0	77.0	0.0
6	4.0	4.0	1000.0	4.0	0.0
7	2.0	7.0	1750.0	14.0	1.0
8	1.0	12.0	3000.0	35.0	0.0
9	2.0	9.0	2250.0	22.0	1.0
10	5.0	46.0	11500.0	98.0	1.0
11	4.0	23.0	5750.0	58.0	0.0
12	0.0	3.0	750.0	4.0	0.0
13	2.0	10.0	2500.0	28.0	1.0
14	1.0	13.0	3250.0	47.0	0.0
15	2.0	6.0	1500.0	15.0	1.0
16	2.0	5.0	1250.0	11.0	1.0
17	2.0	14.0	3500.0	48.0	1.0
18	2.0	15.0	3750.0	49.0	1.0
19	2.0	6.0	1500.0	15.0	1.0
20	2.0	3.0	750.0	4.0	1.0
21	2.0	3.0	750.0	4.0	1.0
22	4.0	11.0	2750.0	28.0	0.0
23	2.0	6.0	1500.0	16.0	1.0
24	2.0	6.0	1500.0	16.0	1.0
25	9.0	9.0	2250.0	16.0	0.0
26	4.0	14.0	3500.0	40.0	0.0
27	4.0	6.0	1500.0	14.0	0.0
28	4.0	12.0	3000.0	34.0	1.0
29	4.0	5.0	1250.0	11.0	1.0

Fig. 1 Blood transfusion dataset

4.1 Bagging

Another name for the categorization algorithm Bagging is Bootstrap Aggregating. In 1994, Leo Breiman made the proposal. The method of choosing samples from the original sample or population and using those samples to estimate various statistics and model correctness is a straightforward yet effective ensemble meta-algorithm. Choosing items with replacements from an original sample N is one method of choosing the bootstrap samples [8]. Model classifier is used on each bootstrap sample, for training or building a model based on the popularity votes [9]. Bagging is used for samples with maximum popularity votes and Regression is used for samples with average popularity votes.

Breiman [10] invented the bagging technique, which is described in the following pseudo code:

Bagging (set D, iterations T, prediction algorithm A) [11]

(1) Model generation: for i = 1 to T, create a bootstrap sample D(i) from D.

Let M(i) represent the result of training A on D(i).
(2) Prediction for i = 1 to T for every test instance w:

Let C(i) be the result of M(i) on w, the most common return class among C(1)… C(T) [11].

4.2 Logistic Regression

Applying well-established Logistic Regression techniques to the AdaBoost approach developed by Jerome Friedman, Trevoe Hastie, and Robert Tibshirani is known as LogitBoost. LogitBoost can be seen as a convex optimization [12]. It is a meta algorithm. Here, an additive model can seek in the form [12]:

$$f = \Sigma_t \alpha_t h_t \tag{1}$$

the LogitBoost algorithm minimizes the logistic loss:

$$\text{logistic loss} = \sum_i \log\left(1 + e^{-y_i f(x_i)}\right) \tag{2}$$

4.3 Classsification ViaRegression

Classification ViaRegression is a tree algorithm [13]. In this case, one can estimate a linear regression function using

$$\begin{aligned} f(x; w) &= W_0 + W_1 X_1 + \ldots + W_d \\ &= W_0 + x^T W_1 \end{aligned} \tag{3}$$

If $y = f(x; w) + !, ! \sim N(0, \sigma2)$, then least squares fitting is the ML objective for parameter w [13]:

$$J_n(w) = \Sigma\left(y_i - f(x_i; w)\right)^2 \tag{4}$$

4.4 J48

J48/C4.5 are CART versions that come after CART and support missing data points in addition to continuous and discrete characteristics. This is ID3's expansion. Ross Quinlan was the developer of this. The algorithm is a tree one. J48 algorithm is the Java version of the C4.5 algorithm. Using the idea of information entropy, this technique creates a decision tree from a collection of training data in a manner similar to ID3 [14]. Figure 2 displays the C4.5 algorithm's pseudo code.

Fig. 2 Pseudo code for J48
algorithm [15]

Algorithm : J48/C4.5(D)

Input: an attribute-valued dataset D

1: Tree = {}

2: if D is "pure" OR other stopping criteria met then

3:      terminate

4: end if

5: for all attribute a € D do

6:        Compute information-theoretic criteria if we split on a

7: end for

8: a_{best} = Best attribute according to above computed criteria

9. Tree = a decision node is created which tests a_{best} in the root of the tree

10: D_x =Induced sub-datasets from D based on a_{best}

11: for all D_x do

12: $Tree_x$ = C4.5(D)

13:   Attach $Tree_x$ to the corresponding branch of Tree

14: end for

15: return Tree

4.5 RandomForest

The technique known as "RandomForests" was expanded by Leo Breiman and Adele
Cutler [12]. Tree learners are trained using the generic bootstrap aggregation technique, also known as bagging, by the random forests training algorithm. The algorithm is a tree one. Using bagging, a random sample is selected and fitted to trees.
This process is repeated (B times) with responses $(Y = y_1,..., y_n)$ and a training set
$(X = x_1,..., x_n)$. Following this process, RandomForest is reached. Another name
for this technique is "feature bagging." Pseudocode for RandomForest [12]:

1. Choose "k" features at random from a total of "m" features, with k < < m.
2. Determine the node "d" among the "k" features by utilizing the optimal split
   point.
3. Using the best split, divide the node into daughter nodes.
4. Continue steps 1 through 3 until "1" nodes are reached.
5. Create a forest by going through steps 1 through 4 "n" times to get "n" trees [12].

5 Experimental Results

The following performance metrics [16] were used to compare the accuracy, sensitivity, specificity and error rate of the various classification algorithms with Blood
Transfusion dataset.

Accuracy: The percentage of true positive and true negative results can be
computed in each analyzed case to determine a test's accuracy. This can be expressed

mathematically as [16]:

$$\text{Accuracy} = \frac{TP + TN}{TP + TN + FP + FN} \tag{5}$$

Sensitivity: The percentage of true positives can be computed to estimate it. This can be expressed mathematically as [16]:

$$\text{Sensitivity} = \frac{TP}{TP + FN} \tag{6}$$

Specificity: The percentage of genuine negative can be computed to estimate it. This can be expressed mathematically as [16]:

$$\text{Specificity} = \frac{TN}{TN + FP} \tag{7}$$

Error Rate: The percentage of false positives and false negatives in each analyzed case can be computed to determine the accuracy of a test. This can be expressed mathematically as follows [16]:

$$\text{Error Rate} = \frac{FP + FN}{TP + TN + FP + FN} \tag{8}$$

where TN-True Negative, FN-False Negative, TP-True Positive and FP-False Positive [16].

5.1 RandomForest

Here, the RandomForest technique is used to classify the Blood Transfusion dataset as well as the normalized Blood Transfusion dataset. Tables 1 and 2 present the findings of a comparative analysis of the accuracy, sensitivity, specificity, and time consumed for each of the two datasets.

Prior to normalization, the RandomForest algorithm had a 93.1% accuracy rate, 93.0% sensitivity, 93.7% specificity, and 6.81% error rate. Following normalization, 93.1% accuracy, 93.1% sensitivity, 93.1% specificity, and 6.81% error rate were attained. Based on the metric values of both datasets, we may infer that the normalized dataset's sensitivity increases by 0.1% and its specificity lowers by 0.6%. Because of this, the Normalized dataset is relatively effective for RandomForest Algorithm.

Table 1 Performance comparison before normalization

Algorithm	Accuracy (%)	Sensitivity (%)	Specificity (%)	Time taken to test model (Sec)
Random forest	93.1	93.0	93.7	0.02
Bagging	82.6	84.0	73.5	0.0
LogitBoost	80.6	82.1	68.9	0.0
ClassificationViaRegression	80.2	82.4	65.3	0.0
J48/C4.5	80.8	85.4	62.0	0.0

Table 2 Performance comparison after normalization

Algorithm	Accuracy (%)	Sensitivity (%)	Specificity (%)	Time taken to test model (Sec)
Random forest	93.1	93.1	93.1	0.03
Bagging	82.4	83.7	74.2	0.0
LogitBoost	80.6	82.1	68.9	0.0
ClassificationViaRegression	80.2	82.4	65.3	0.0
J48/C4.5	80.8	85.4	62.0	0.0

5.2 Bagging

Here, the Blood transfusion dataset and the normalized Blood transfusion dataset are classified using Bagging Algorithm. The accuracy, sensitivity, specificity, time taken for both the datasets are studied comparatively and Tables 1 and 2 tabulate the corresponding results.

The Bagging algorithm has the accuracy of 82.6%, sensitivity of 84.0%, specificity of 73.5% and error rate of 17.3% before normalization. After normalization, the Bagging algorithm has the accuracy of 82.4%, sensitivity of 83.7%, specificity of 74.2% and error rate of 17.5%. From the metrics values we conclude that the value of accuracy decreases by 0.2% and sensitivity decreases by 0.3% and specificity increases by 0.7% and error rate increases by 0.2% after normalization. Therefore, for Bagging algorithm the unnormalized dataset is comparatively efficient.

5.3 LogitBoost

In this, the Blood transfusion dataset and the normalized Blood transfusion dataset are classified using LogitBoost Algorithm. The accuracy, sensitivity, specificity, time taken for both the datasets are studied comparatively and Tables 1 and 2 tabulate the corresponding results.

The LogitBoost algorithm has the accuracy of 80.6%, sensitivity of 82.1%, specificity of 68.9% and error rate of 19.3% before normalization. After normalization, this algorithm achieves the accuracy of 80.6%, sensitivity of 82.1%, specificity of 68.9% and error rate of 19.3%. From the metrics value we observe that there is change in the metric values. Therefore for LogitBoost algorithm the metrics value does not vary due to normalization (i.e.) normalization has no effect on LogitBoost Algorithm.

5.4 ClassificationViaRegression

Here, the Blood transfusion dataset and the normalized Blood transfusion dataset are classified using ClassificationViaRegression Algorithm. The accuracy, sensitivity, specificity, time taken for both the datasets are studied comparatively and Tables 1 and 2 tabulate the corresponding results.

The ClassificationViaRegression algorithm has the accuracy of 80.2%, sensitivity of 82.4%, specificity of 65.3% and error rate of 19.7% before normalization. After normalization, this algorithm achieves the accuracy of 80.2%, sensitivity of 82.4%, specificity of 65.3% and error rate of 19.7%. From the metrics value we observe that there is change in the metric values. Therefore for Classification ViaRegression algorithm the metrics value does not vary due to normalization (i.e.) normalization has no effect on Classification ViaRegression Algorithm.

5.5 J48

In this, the Blood Transfusion dataset and the normalized Blood Transfusion dataset are classified using J48 Algorithm. The accuracy, sensitivity, specificity, time taken for both the datasets are studied comparatively and Tables 1 and 2 tabulate the corresponding results.

The J48 algorithm has the accuracy of 80.8%, sensitivity of 85.4%, specificity of 62.0% and error rate of 19.1% before normalization. After normalization, this algorithm achieves the accuracy of 80.8%, sensitivity of 85.4%, specificity of 62.0% and error rate of 19.1%. From the metrics value we observe that there is change in the metric values. Therefore for J48 algorithm the metrics value does not vary due to normalization (i.e.) normalization has no effect on J48 Algorithm.

5.6 *Performance Comparison*

The performance of all classification algorithms was compared with accuracy, sensitivity, specificity and time taken to test model. This is shown in Tables 1 and 2..

Figure 3 shows the value of error rate of various algorithms before normalization and have come to a conclusion that RandomForest is the best algorithm as it has the lowest value of 6.81% compared to all the other algorithms.

Figure 4 shows the value of error rate after normalization and has concluded that the RandomForest algorithm is the best algorithm as it has the lowest error rate of 6.81%.

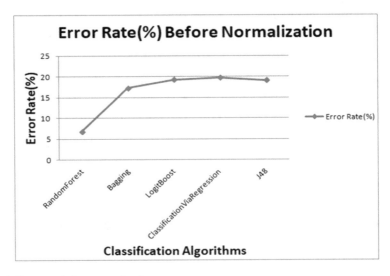

Fig. 3 Error rate before normalization

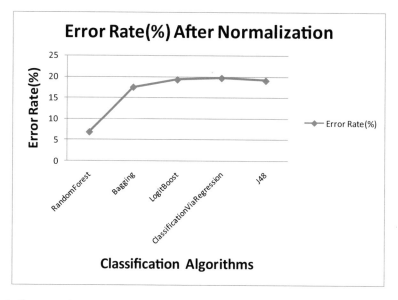

Fig. 4 Error rate after normalization

6 Conclusion

Decision trees are often non parametric, meaning they don't assume anything about the data's distribution. With the Blood transfusion dataset, the RandomForest algorithm outperforms other algorithms in terms of accuracy and error rate. The optimal algorithm selected is RandomForest; nevertheless, the un-normalized dataset takes 0.02 s to process, while the normalized dataset takes 0.03 s. Normalization is therefore not always required, but depending on the algorithm, decisions about whether to do so or not can be made. In the future, we can compare the complexity of each classification algorithm.

References

1. Blood Transfusion Service Center Home Page (2023) https://archive.ics.uci.edu/dataset/176/blood+transfusion+service+center. Accessed 24 Oct 2023
2. Dimov R, Feld M, Dr Kipp M, Dr Ndiaye A, Dr Heckmann D (2007) Weka: practical machine learning tools and techniques with java implementations. AI Tools Semin Univ Saarl, WS 6(07)
3. Santhanam T, Sundaram S (2010) Application of CART algorithm in blood donors classification. J Comput Sci 6(5):548
4. Sundaram S (2011) A comparison of blood donor classification data mining models. J Theor Appl Inf Technol 30(2)
5. Darwiche M (2010) Prediction of blood transfusion donation. Research challenges in information science (RCIS). In 2010 Fourth International conference on, IEEE

6. Zabihi F, Ramezan M, Pedram MM, Memariani A (2011) Rule extraction for blood donators with fuzzy sequential pattern mining. J Math Comput Sci 2(1)
7. Lee WC, Cheng BW (2011) An intelligent system for improving performance of blood donation. 18(2):173–185
8. Anderson B (2019) Pattern recognition: an introduction. Scientific e-Resources
9. Seni G, Elder J (2010) Ensemble methods in data mining: improving accuracy through combining predictions. Morgan & Claypool Publishers
10. Shona D, Senthilkumar M (2016) An ensemble data preprocessing approach for intrusion detection system using variant firefly and Bk-NN techniques. Infinite Study
11. Machine Learning Home Page (2023) http://www.cs.bc.edu/~alvarez/ML/bagging.html. Accessed 27 Oct 2023
12. Schapire RE, Freund Y (2013) Boosting: foundations and algorithms. Kybernetes 42(1):164–166
13. Conati C, McCoy K, Paliouras G (eds) User modeling 2007: 11th international conference, UM 2007, Corfu, Greece, July 25–29, 2007. Proceedings (vol 4511). Springer Science & Business Media
14. Larose DT, Larose CD (2014) Discovering knowledge in data: an introduction to data mining, vol 4. John Wiley & Sons
15. Wu X, Kumar V (eds) The top ten algorithms in data mining. CRC press
16. Han J, Kamber M, Pei J Data mining concepts and techniques third edition. University of Illinois at Urbana-Champaign Micheline Kamber Jian Pei Simon Fraser University

Water Body Segmentation for Satellite Images Using U-Net++

G. Rajalaxmi, S. E. Vimal, and Janani Selvaraj

Abstract Satellite images are important for both m onitoring and managing natural resources. The ability to identify and manage water resources is made possible by the segmentation of water bodies in satellite data. In this study, U-Net++ (Nested U-Net) model was used to separate water bodies in satellite data. The dataset for the project was collected using USGS Earth Explorer and QGIS, and it was divided into 20% for testing and 80% for training. After 70 cycles of training, the U-Net++ model had an accuracy of 97.66%. The U-Net++ model builds on the original U-Net model, which has been widely used for segmentation tasks. The U-Net++ model incorporates skip connections and dense connections to improve model performance. This study's ability to segment the water body opens up a lot of possibilities for controlling and monitoring water supplies, among other things. The accuracy reached with the U-Net++ model demonstrates its capacity for accurate water body segmentation in satellite pictures.

Keywords U-Net++ · QGIS · Satellite imagery · USGS

1 Introduction

Practically every part of our lives, including the air, the land, and the sea, are all traversed by water, which modifies each as it passes. It is a resource that is vital to the world, human existence, and business operations. Human activity and climate change in their geographical and historical contexts Due to changes in the distribution of water bodies, there are currently fewer water resources that may be used. The state of the globe is slowly changing. So, one use of technology is to understand the location and quantity of water on the Earth's surface to minimize this problem. One of the more interesting new study fields for research is the use of remote sensing in water level detection. using remote sensing, a method of detecting and monitoring a location's

G. Rajalaxmi (✉) · S. E. Vimal · J. Selvaraj
Department of Data Science, Bishop Heber College, Tiruchirappalli, Tamil Nadu, India
e-mail: rajiganesan1999@gmail.com

© The Author(s), under exclusive license to Springer Nature Singapore Pte Ltd. 2024
D. Giri et al. (eds.), *Proceedings of the Tenth International Conference on Mathematics and Computing*, Lecture Notes in Networks and Systems 963,
https://doi.org/10.1007/978-981-97-2069-9_7

81

physical features by seeing its emitted radiation from a distance, to analyze the data by Active and passive approaches make up the two categories of remote sensing. When an object's location, velocity, and direction are calculated from measurements of the time between emission and return, this is referred to as "actively" remotely sensed data. Radiation that is emitted or reflected by the item or its surroundings is collected by passive sensors. The most frequent form of radiation detected by passive sensors is reflected sunlight.

The aim to extract data characteristics from satellite pictures using remote sensing methods and deep learning algorithms is to detect surface water by using the concepts of convolutional neural networks (CNNs). It is a unique type of network design for deep learning algorithms that is useful for tasks like image identification, picture segmentation, and pixel data processing. Every pixel in a raw picture is given a preset class name through the process of segmentation. Water-body extraction is a common use for automatic segmentation, which is a core component of remote sensing and picture interpretation. The convolutional neural network known as Nested U-Net has been extensively used for image segmentation in the segmentation of satellite pictures. Nested U-Net is widely used because of its qualities, such as fast and accurate photo segmentation. In this network, the features are first recovered using downsampling, and then segmented masks of the image are produced using upsampling. Because it is supervised, this model requires data to be trained. To be segmented during testing in the actual world, these networks train on images and the corresponding masks, learning the characteristics that exist inside the mask. Calculating the water body's area is the major objective of this project, which also aims to monitor water bodies and do hydrological research.

2 Related Work

Miao et al. [1], proposed three networks for DeconvNet: a convolution network, a deconvolution network, and a restricted receptive field (RRF) deconvolution network. The convolution network combines convolution and pooling layers with residual units, while skip connections between convolution and deconvolution layers enhance detail preservation. Despite its success, the model is outdated for segmenting satellite images.

Yuan et al. [2], introduced a methodology using the Deeplabv3 model for segmenting remote sensing data. They addressed limitations of the Deeplabv1 architecture by proposing Deeplab V2, which incorporates atrous separable convolution for improved performance. However, Deeplab models are considered outdated for training on satellite images.

Rama Charan et al. [3], employed machine learning techniques to extract water resources from non-water bodies using BHUVAN open data. They emphasized the use of CNN and introduced Rasterio, a Python library, for efficient geographic raster data handling.

Muhadi et al. [4], compared DeepLabv3+ and SegNet for water body segmentation. DeepLabv3+ outperformed SegNet, demonstrating potential for water monitoring and flood control.

Chatterjee et al. [5], used the ensemble Detectron2 instance segmentation architecture to monitor significant changes in geographical features using satellite photos. Mask R-CNN with ResNet50+ FPN backbone showed the best performance.

Bhangale et al. [6], discussed the use of the normalized difference water index (NDWI) for water body segmentation in satellite photography. They highlighted the effectiveness of NDWI across various satellite images.

Rana et al. [7], emphasized the importance of water and proposed the use of conventional supervised maximum likelihood classification for data segmentation. The approach involves training a model on labeled data to predict class labels for new, unlabeled data.

Yulianto et al. [8], integrated the DeepLabv3+ network to improve water body extraction over large distances. Their method demonstrated high prediction accuracy, outperforming other semantic segmentation approaches.

Chatterjee1 et al. [9], recommended Kapur's entropy-based thresholding for aquatic organism segmentation from very high-resolution (VHR) satellite photos. The proposed approach showed effectiveness in segmentation.

Zaffaroni et al. [10], compared SegNet and UNet, both based on ResNet, for water segmentation. SegNet outperformed other algorithms in terms of prediction speed, memory use, VRAM, and size.

Concluding the Related Work section, it is observed that various approaches have been proposed for water body segmentation in satellite images, with each method having its strengths and limitations. Notably, the choice of model and techniques depends on specific application requirements, such as accuracy and prediction speed.

3 Methodology

3.1 Dataset Description

The two folders that make up the dataset are the Image and Mask directories in the water body directories. Between the two files, there are 5000 images. The image folder includes 148 × 148 pixel resolution RGB images of satellite water pictures. The mask folder contains the 148 × 148-pixel grayscale images of the masked satellite water images. A number of water body pictures are selected from the USGS Earth Explorer and QGIS. Each image has a black-and-white mask, in which the color black stands for anything other than water and the color white for water. The CV2 custom threshold technique, which is frequently used to recognize and quantify vegetation in satellite images, was utilized to construct the masks. To detect aquatic bodies, a higher threshold was used (Fig. 1).

Fig. 1 Water body images
and their respective masks

SATELLITE IMAGE   MASK IMAGE

3.2   Methods of Collecting Data

The primary goal of this research was to acquire enough data to train a U-Net++ model. Unfortunately, gathering and annotating data is one of the most common obstacles in data science and machine learning projects. To create own dataset, obtained the USGS's geographic information and entered it into USGS Earth Explorer. This online platform provides access to a variety of geospatial and remote sensing data, including topographic maps, satellite images, and aerial photographs. The geographic information was downloaded in the shapefile format. A shapefile is a popular geographic vector data format used in Geographic Information Systems (GIS) applications, developed and maintained by Esri. QGIS is a free and open-source geographic information system (GIS) application that is produced by a team of volunteers and works with Windows, Mac, and Linux operating systems. It provides a range of features and capabilities for working with geographic data, including the ability to design, edit, visualize, analyze, and publish maps. It also features built-in support for popular open data sources like OpenStreetMap. QGIS's main characteristics include the following: Many projections and kinds of geographic data are supported; network analysis, interpolation, and geoprocessing plugins provide additional functionality; and integration support is provided for more open-source programs like R and GRASS GIS. It is regarded as one of the most robust and user-friendly GIS tools on the market. After downloading the satellite images, used binary threshold method to create an mask image.

The binary thresholding approach is used to turn a grayscale picture into a binary image. It involves choosing a threshold value and comparing each pixel in the picture to it. Many methods, such as automatic thresholding algorithms and human thresholding, can be used to select the threshold value. Binary threshold method is the most popular thresholding technique, which selects the threshold value based on the image's histogram. Binary thresholding is widely used as a pre-processing step when carrying out tasks like picture segmentation, object identification, or feature extraction. However, if the threshold value is chosen improperly, the image has noise, or the lighting shifts, it can result in information loss.

3.3 Preprocessing

Preprocessing methods play a significant role in getting image data ready for deep learning models. Lots of preprocessing techniques were used to prepare own datasets. By isolating the relevant areas of a picture by cropping, models may more easily learn from those areas. Another often used method is resizing, which enables uniformly sized photos or reduces computing complexity. Another method of lowering the number of color channels and making the data simpler for the model is to convert RGB photos to grayscale. Images may be turned into binary images via binary thresholding, A popular image processing approach in computer vision and image analysis is the binary threshold method. Setting a threshold value for an image's pixel intensities entails assigning all pixels above the threshold to a maximum value (often 255) and any pixels below the threshold to a minimum value (usually 0). A binary picture is the end result, with each pixel being either wholly black or entirely white. The binary threshold approach may be used to differentiate water pixels from non-water pixels in an image when segmenting water bodies in satellite pictures. As water pixels often have lower pixel intensities than non- water pixels, a suitable threshold may be used to differentiate between these two kinds of pixels. Pixels can be classified as being either water or non-water depending on whether their intensities are below or over the threshold. further Normalization, which equalises the impacts of pixel values on the model by scaling them to a common range, is another efficacy metric. The image and mask data are kept in turn in two arrays called train images and train masks.

Each train picture element is divided by 255 to produce a new array x with normalized pixel values between 0 and 1. In a manner similar to this, train masks/255 divides each element of train masks by 255 to produce a new array y that includes normalized mask values between 0 and 1. By normalizing the pixel values, the model-training process may converge more rapidly and become more stable. Moreover, issues like vanishing or inflating gradients may occur when the range of values is too large, which may be prevented by normalizing.

Finally, made a data augmentation, for the purpose of training a model, sets of photographs are generated using an image data generator, a deep learning method. To better utilise RAM, it imports the pictures in groups rather than importing the entire dataset at once. Other methods of data augmentation, such as random rotations, zooms, and flips, can also be used by image data producers to produce new images and increase the training collection. This improves the model's generalizability and avoids overfitting. To further lessen overfitting, the generator can also mix the data between epochs. To make it simple to build picture data generators for deep learning model training, the Keras framework includes an ImageDataGenerator class.

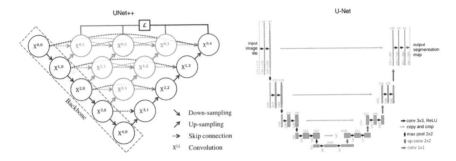

Fig. 2 U-Net++ architecture

3.4   Model Development

The model is developed using U-Net++. It can be a helpful tool for water body segmentation in satellite images because of its wide and symmetric encoder-decoder structure with skip links (Fig. 2).

This approach enables the model to capture the high-level characteristics of the satellite image while preserving the spatial information of the water body pixels. By training the UNet++ model using a sizable dataset of annotated satellite photos and making use of data augmentation techniques, it is possible to teach it to precisely distinguish water bodies in satellite images. The trained model may then be used to infer the freshly obtained satellite images by automatically identifying and separating the pixels that indicate water bodies.

3.5   Model Architecture

The Fig. 3. demonstrates the work flow of the model. This contains multi-stage process for analyzing satellite images and determining how much land and water are included in the image. Each step operates as follows:

(1)   The model accepts both RGB satellite images and binary threshold images as inputs. The RGB shots of the structures, roads, and plants are indicated in the binary images as distinct traits or objects. The characteristics of the earth's surface are shown by the RGB photographs.

(2)   Preprocessing is carried out to prepare the input pictures for analysis. This includes procedures to improve the image quality as well as cropping and resizing the photographs to a standard size, augmentation of the data to increase the amount of data available for training, and other activities.

(3)   The preprocessed pictures are then sent into a UNet++ model to extract features. For tasks involving segmenting or separating an image into regions or segments

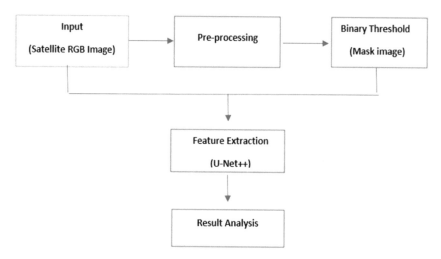

Fig. 3 Work flow

depending on particular attributes, the UNet++ model is a well-known deep learning model.

(4) The last step in estimating the amount of land and water in the satellite image is the output analysis of the UNet++ model. By counting the number of pixels in each segment or region or by using alternative techniques, it is possible to determine the total area of land and water in the image.

3.6 Web-Application

Web application has created as part of the project that would let users to submit satellite images and use the U-Net++ model to locate water bodies in those images. Front-end is created using CSS and HTML to make the application simple and straightforward to use. The front interface of the program and the U-Net++ model were combined using the Flask framework. The popular Python web application framework Flask can be used to build web apps and connect them to databases and back-end models. Users can now upload images and obtain real-time results for water body segmentation.

4 Results

For the segmentation of water bodies in satellite photos, use U-Net and U-Net++. My result showed that after 70 iterations, the U-Net++ model had a substantially greater accuracy of 90%, whereas the U-Net model had just 75% accuracy. These findings imply that the U-Net++ model is superior to the U-Net model in identifying and classifying water features in satellite imagery. The U-Net++ model's design, which is an expansion of the U-Net model and incorporates extra skip connections that assist in retaining spatial information and enhance feature representation, may be responsible for the model's excellent.

Accuracy. Keep in mind that the accuracy of the model may vary depending on the specifics of the input data. Once the model is trained, use any water body picture submitted by the user in the UI to train and test masked and unmasked images to determine the percentage of water present in the masked image. The masked image distinguishes between the land and the water by using white pixels to represent the water's edge and black pixels to represent the remainder of the image. A useful outcome of this is the computation of the water body's area using drones to ascertain, among other things, whether the area is at risk of flooding or cannot be used for human cultivation (Figs. 4, 5 and 6).

A web application was created to compare real-time analysis of the most recent image with earlier ones and identify the region of difference. An interactive user interface appears as soon as the web programme is opened. The user can upload any image by selecting it from the UI and hitting the "Select Image" button. Once everything is in position, the user simply needs to press the submit button, and the outcome will appear on his screen shortly after (Fig. 7).

MODEL	ACCURACY
U-NET++	97%
U-NET	75%

Fig. 4 Result analysis

```
accuracy
        training          (min:   0.235, max:   0.985, cur:   0.980)
        validation        (min:   0.208, max:   0.980, cur:   0.977)
Loss
        training          (min:   0.400, max:   1.803, cur:   0.405)
        validation        (min:   0.391, max: 151.047, cur:   0.412)

Epoch 70: saving model to satellite_unet.hdf5
14/14 [==============================] - 2s 168ms/step - loss: 0.4054 - accuracy: 0.9801 - val_loss:
0.4117 - val_accuracy: 0.9766 - lr: 1.0000e-05
```

Fig. 5 Model results

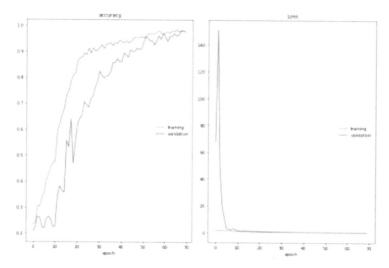

Fig. 6 Result accuracy and loss plots

Results:	
Water Percentage	25.46%
Land Percentage	74.54%

Fig. 7 Final output

5 Conclusions

Using a neural network, more specifically a convolutional neural network, pictures were successfully classified into categories of either water bodies or non-water bodies (CNN). Users may upload any RGB satellite image using this method into the newly created user interface system. By examining the reflecting properties of water bodies, the computer can determine how much water is in a given image. CNNs are particularly suitable for use in image classification problems as they can efficiently find patterns and attributes within an image that are relevant to the classification aim. In this situation, the CNN was trained to recognise the specific visual characteristics that distinguish water bodies from other types of objects in satellite pictures. As the model takes advantage of how water bodies reflect light, it can classify water bodies accurately even when they are partially obscured or hidden by other features.

Overall, this approach provides a powerful tool for analysing satellite images of water bodies, which has numerous uses in resource management, disaster response, and environmental monitoring. By allowing people to contribute their own photos and gauge the water content, this technology has the potential to enable a wide range of applications in both academic and practical contexts.

References

1. Miao Z, Fu K, Sun H, Sun X, Yan M (2020) Automatic water-body segmentation from high-resolution satellite images via deep networks. IEEE Geosci Remote Sens Lett
2. Yuan K, Zhuang X, Schaefer G, Feng J, Guan L, Fang H (2021) Deep-learning-based multispectral satellite image segmentation for water body detection. IEEE
3. Rama Charan DL, Sai Siva Teja D, Subhashini R, Bevish Jinila Y, Gandhi M (2020) Convolutional neural network based water resource monitoring using satellite images. IEEE
4. Muhadi NA, Abdullah AF, Bejo SK1, Mahadi MR1, Mijic A (2021) Deep learning semantic segmentation for water level estimation using surveillance camera. MDPI
5. Chatterjee R, Chatterjee A, Hafzul Islam3 SK (2021) Deep learning techniques for observing the impact of the global warming from satellite images of water-bodies. Springer
6. Yulianto F, Kushardono D, Budhiman S, Nugroho G, Chulafak GA, Dewi EK, Pambudi AI (2022) Evaluation of the threshold for an improved surface water extraction index using optical remote sensing data. Sci World J
7. Zaffaroni M (2018) Water segmentation with deep learning models for flood detection and monitoring. Zaffaroni et al. 2018
8. Aalan Babu A, Mary Anita Rajam V (2020) Water-body segmentation from satellite images using Kapur's entropy-based thresholding method
9. Bhangale U, More S, Shaikh T (2020) Analysis of surface water resources using sentinel-2 imagery. In: Third international conference on computing and network communications (CoCoNet'19)
10. Li J, Ma R, Cao Z (2020) Satellite detection of surface water extent: a review of methodology. MDPI
11. Wang X, Xie H (2018) A review on applications of remote sensing and geographic information systems (GIS) in water resources and flood risk management. Water 10
12. Shao Z, Fu H, Li D, Altan O, Cheng T (2019) Remote sensing monitoring of multi-scale watersheds impermeability for urban hydrological evaluation. Remote Sens Environ 232
13. Wurm M, Stark T, Zhu XX, Weigand M, Taubenböck H (2019) Semantic segmentation of slums in satellite images using transfer learning on fully convolutional neural networks. ISPRS J Photogramm Remote Sens 150:59–69
14. Sagin J, Sizo A, Wheater H, Jardine TD, Lindenschmidt K-E (2015) A water coverage extraction approach to track inundation in the Saskatchewan River Delta, Canada. Int J Remote Sens 36(3):764–781. https://doi.org/10.1080/01431161.2014.1001084
15. Mueller N, Lewis A, Roberts D et al (2016) Water observations from space: mapping surface water from 25 years of landsat imagery across Australia. Remote Sens Environ 174:341–352. https://doi.org/10.1016/j.rse.2015.11.003.-
16. Topp SN, Pavelsky TM, Jensen D, Simard M, Ross MRV (2020) Research trends in the use of remote sensing for inland water quality science: moving towards multidisciplinary applications. Water 12(1):169–234
17. Li L, Yan Z, Shen Q, Cheng G, Gao L, Zhang B (2019) Water body extraction from very high spatial resolution remote sensing data based on fully convolutional networks, environmental science, mathematics. Remote Sens
18. Dang B, Li YS (2021) MSResNet: multiscale residual network via self-supervised learning for water-body detection in remote sensing imagery. Remote Sens 13:3122
19. Danesh-Yazdi M, Bayati M, Tajrishy M, Chehrenegar B (2021) Revisiting bathymetry dynamics in Lake Urmia using extensive field data and high-resolution satellite imagery. J Hydrol 603:17

20. Li D, Wu B, Chen B, Xue Y, Zhang Y (2020) Review of water body information extraction based on satellite remote sensing. J Tsinghua Univ Sci Technol 60:147–161
21. Zhang D, Yang S, Wang Y, Zheng W (2019) Refined water body information extraction of Three Gorges reservoir by using GF-1 satellite data. Yangtze River 50:233–239
22. He H, Huang X, Li H (2020) Water body extraction of high resolution remote sensing image based on improved U-Net network. J Geo-Inf Sci 22

ResNet-CPDS: Colonoscopy Polyp Detection and Segmentation Using Modified ResNet101V2

S. K. Hafizul Islam, **Jitesh Pradhan**, **Purnendu Vashistha**, **Aman P. Singh**, and **Aman Kishore**

Abstract Colorectal cancer (CRC) is a global public health concern, and early detection through screening reduces mortality rates. It is one of the common types of cancer with a high mortality rate. Traditionally, colonoscopy is used to detect CRC, which is inefficient. Therefore, an automated Colonoscopy Polyp Detection and Segmentation (CPDS) system can significantly increase the efficiency of colonoscopy. We propose an automated model, ResNet-CPDS, using the modified ResNet101V2 model. We evaluate the performance of ResNet-CPDS and other CPDS models, and compare their accuracy. We also demonstrate that the ResNet-CPDS model outperforms other models for the CVC-ClinicDB dataset.

Keywords Colorectal cancer · Polyp segmentation and detection · Precision · Recall · ResNet101V2

S. K. H. Islam · P. Vashistha · A. P. Singh · A. Kishore
Department of Computer Science and Engineering, Indian Institute of Information Technology Kalyani, West Bengal 741235, India
e-mail: hafi786@gmail.com

P. Vashistha
e-mail: mailtopv2299@gmail.com

A. P. Singh
e-mail: aman.pratap.7389@gmail.com

A. Kishore
e-mail: amankishorepaul@gmail.com

J. Pradhan (✉)
Department of Computer Science and Engineering, National Institute of Technology Jamshedpur, Jharkhand 831014, India
e-mail: jiteshpradhan.cse@nitjsr.ac.in

© The Author(s), under exclusive license to Springer Nature Singapore Pte Ltd. 2024
D. Giri et al. (eds.), *Proceedings of the Tenth International Conference on Mathematics and Computing*, Lecture Notes in Networks and Systems 963,
https://doi.org/10.1007/978-981-97-2069-9_8

93

1 Introduction

Colorectal cancer (CRC) is a large group of diseases that can start in almost any organ or tissue of the body. It occurs when abnormal cells grow out of control, cross their normal boundaries to infect nearby body parts, and spread to other organs. The latter process, known as *metastasizing*, significantly contributes to cancer-related mortality. Cancer is the second leading cause of death globally, and, in 2018, there were approximately 18 million cases globally, of which 1.5 million were in India alone [1]. India had around 0.8 million cancer deaths in 2018, against 9.5 million globally. The number of new cases is estimated to double in India by 2040. Out of the different causes of cancer, CRC is one of them. Most CRCs start as a growth on the inner lining of the colorectal or rectum. These growths are called polyps that generally turn into cancer.

1.1 Related Works

In the literature, many Convolution Neural Network (CNN)-based models were adopted to address the problem of Colonoscopy Polyp Detection and Segmentation (CPDS). Chatterjee et al. [2] proposed a Trident U-Net model for CPDS, combining the encoders of ResNet50, MobileNetV2, and EfficientNetB0. The performance of the model was evaluated on CVC-612 and ETIS datasets based on the metrics: Dice Similarity Coefficient (DSC), Intersection-over-Union (IoU) score, Recall, and Precision. On the CVC-612 dataset, the Trident model achieved a DSC of 90%, IoU score of 82%, precision of 73.61%, and recall of 86.31%. This model exhibits a quicker convergence rate but requires longer training due to its larger number of parameters. Mo et al. [3] proposed a Fast R-CNN with a VGG16 model as a backbone. They used the CVC15 test dataset and achieved a precision of 97.2%, recall of 85.2%, F_1-score of 90.8%, and F_2-score of 87.4%. For the CVC15 training dataset, they achieved a precision of 86.2%, recall of 98.1%, F_1-score of 91.7%, and F_2-score of 95.6%. However, in their model, a high false-positive rate is observed during localization tests, especially in frames with polyps. Based on ResNet101, Jha et al. [4] proposed ColonSegNet that achieved an average precision of 0.8745, and YOLOv4 (Darknet53) that achieved real-time speed (48FPS) with an average precision of 0.8513. ResNet-34 was used for segmentation, while the ColonSegNet achieved competitive scores with faster processing. Shruti et al. [5] evaluated different models on the Kvasir-SEG dataset with ResNet-34 and EfficientNetB2 as their encoder backbones. They used Tversky loss, performed CutMix, and standard augmentations for data preprocessing, and achieved the IoU of 75.50% with an accuracy of 95.83%.

1.2 Motivations and Contributions

CPDS is a critical process in identifying and outlining polyps in medical images, and it plays a vital role in the early detection of CRC. However, this task presents significant challenges due to the absence of well-defined edges between the affected region and its surroundings. Polyps exhibit a wide range of shapes, sizes, colors, and textures, making their recognition even more intricate. Polyps are also minuscule and challenging to discern, particularly in low-resolution images. Furthermore, the scarcity of datasets specifically curated for training models in polyp segmentation further compounds the difficulty in developing accurate and robust models for this task. The combination of these factors underscores the complexities associated with polyp segmentation and emphasizes the need for innovative techniques and methodologies to overcome these challenges.

In this study, we investigated CNN-based architectures for the development of an automated CPDS system. Specifically, we focused on the ResNet and efficient models. Our findings revealed that the ResNet101V2 model displayed promising performance, achieving high accuracy, recall, precision, and IoU scores. To further improve its performance, we identified potential strategies such as increasing training epochs, applying data augmentation techniques, adjusting convolutional filters, and fine-tuning hyperparameters. We modified the ResNet101V2 model, and proposed an improved system, called ResNet-CPDS, for CPDS. Overcoming the challenge of the limited availability of quality datasets, we used CVC-ClinicDB [6] dataset for the evaluation. The outcomes demonstrated that the performance of the ResNet-CPDS model is comparable to state-of-the-art CPDS models.

1.3 Paper Organization

The rest of the paper is organized as follows. Section 2 discussed the various deep learning models used for CPDS. Section 3 discussed the proposed methodology. Section 4 discusses the datasets, experimental setup, results, and comparative. Section 5 concludes the paper with valuable remarks and some future scopes.

2 Preliminaries

Image segmentation involves partitioning an image into multiple regions. There are several different approaches to image segmentation, each with its own advantages and limitations, including thresholding, edge detection, region growing, clustering, and deep learning. The choice of segmentation technique depends on the specific application and the characteristics of the input image. However, CNN-based segmentation networks have shown promising results in recent years, with the U-Net architecture being one of the most popular choices. Now, we will discuss the different variants of U-Net architecture.

- **U-Net Model**: The U-Net [7], U-Net++ [8], and U-Net 3+ [9] architectures are widely adopted in medical image segmentation tasks, which exhibits the capability to capture both local and global features of input images. U-Net implementation is straightforward and can be trained effectively even with limited data. The U-Net++ incorporates nested convolutions, dense skip connections, and deep supervision. However, U-Net 3+ model integrates dense blocks between encoder and decoder paths. Dense blocks consist of densely connected convolutional layers, which promote feature reuse and help capture long-range dependencies in the input. This increases understanding of complex patterns and segmentation accuracy.
- **Trident Model**: Trident [2] is an extension of the U-Net architecture. It was designed to address the limitations associated with multi-scale feature handling. This architecture comprises three parallel U-Net sub-networks, each with a distinct receptive field, allowing the network to capture local and global features at multiple scales. The output of each sub-network is concatenated to generate the final segmentation mask. Trident has demonstrated promising results in CPDS with recall and precision values of 86% and 96%, respectively. A notable advantage of this model is its effective handling of multi-scale features, leading to enhanced segmentation accuracy. Furthermore, Trident can be trained using the same strategies as U-Net, including supervised, semi-supervised, and weakly supervised learning. U-Net and Trident are prominent CNN architectures employed in CPDS. While U-Net has established its efficacy in numerous studies, Trident offers superior performance in handling multi-scale features.

3   Proposed ResNet-CPDS System

We used the ResNet101V2 model to develop an efficient automated CPDS system for assisting medical professionals in identifying polyps at an early stage. The proposed ResNet-CPDS model is depicted in Fig. 1. This model is inspired by the U-Net and Trident architectures. Initially, the dataset is loaded, comprising input images and corresponding labels. Preprocessing techniques, such as resizing, normalization, or data augmentation, are applied to enhance performance. The model specifies the architecture and layers, including convolutional, pooling, and fully connected layers. After compilation, where the optimizer, loss function, and evaluation metrics are set, the model is trained using the fit method. The dataset is iteratively fed to the model for a specified number of epochs to adjust weights and biases and minimize loss. Finally, the trained model enables segmentation by predicting pixel-wise labels or generating bounding boxes to identify and delineate specific objects or regions within an image. The training process (see Fig. 2) of the ResNet-CPDS model is illustrated below.

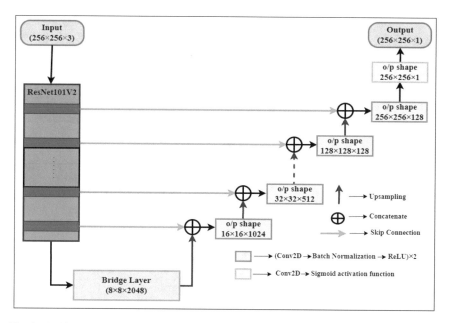

Fig. 1 Architecture of the ResNet-CPDS model

- **Data Preprocessing**: Initially, we cleaned and preprocessed the data to ensure its quality and compatibility with the model. This involved removing errors and inconsistencies and transforming categorical variables into numerical formats. Additionally, we standardized the data by scaling values to a common range, facilitating improved predictions by the models. Lastly, we partitioned the data into training and testing sets to assess the performance of the models on unseen data.
- **Loading Data**: We separated the images and masks from the dataset; sorted them based on their names; and split the data into three parts for training, validation, and testing. This helps to prepare the data for use in the ResNet-CPDS model properly.
- **Resizing and Parsing**: After collecting the dataset, we prepare it for the model. This involves resizing the images and masks to a specific size, such as 256×256 pixels with three color channels for images and 256×256 pixels with a grayscale channel for masks. We ensure the proper pairing of each image and mask by aligning them in the dataset. To enhance training efficiency, we divide the dataset into batches of 16 for more effective processing by the model. This optimization aids in better learning from the data and improves the accuracy of the model's results.
- **Model Training**: In the training phase, we utilized the Adam optimizer, a stochastic gradient descent method. We evaluated the performance using recall and precision metrics, where recall measures the model's ability to detect positive samples and precision measures the accuracy of positive sample classification. We

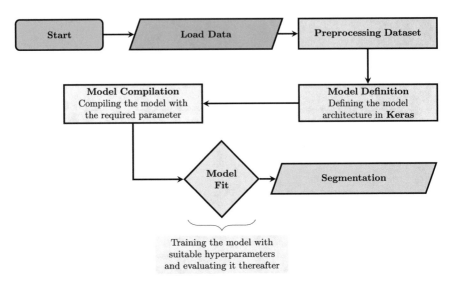

Fig. 2 Flowchart of the training of the ResNet-CPDS model

employed the ReduceLROnPlateau technique to adjust the learning rate dynamically and EarlyStopping to halt training if metrics did not improve within ten epochs. Initially, we took the learning rate as 10^{-4}, but due to ReduceLROnPlateau, it was reduced till 10^{-10}. Different models were trained for different numbers of epochs due to the implementation of EarlyStopping. However, ResNet101V2 ended its training at 24th epoch. The binary cross-entropy loss was used for training.

4 Results and Discussions

This section provides a concise overview of the dataset employed in the experimental analysis. We also present a comprehensive analysis of the experimental results, including comparisons with other state-of-the-art CPDS methods, followed by an in-depth discussion of the findings. Several publicly available datasets have been created to facilitate the development of automated CPDS algorithms. These datasets contain images of polyps from different manufacturers' colonoscopes and ground truth masks that indicate the location and extent of each polyp. For CPDS, we used CVC-ClinicDB dataset [6], comprising 612 images extracted from colonoscopy videos. Ground truth masks corresponding to the polyp regions were provided. The size of each image is 384×288 during data preprocessing, and it changed to 256×256 to train the proposed model. Figure 3 contains some sample images from the CVC-ClinicDB dataset.

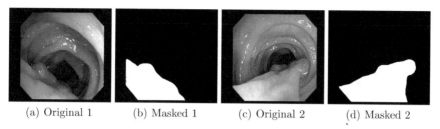

(a) Original 1   (b) Masked 1   (c) Original 2   (d) Masked 2

Fig. 3 Examples from CVC-ClinicDB dataset

We used the following metrics: IoU, DSC, Precision, and Recall [2] to evaluate the combination of the following models: (i) ResNet101 and EfficientNetB3, (ii) ResNet101 and EfficientNetB6, (iii) ResNet152V2 and EfficientNetB7, (iv) ResNet50 with freezing, (iv) ResNet50 without freezing, (vi) EfficientNetB7, and (vii) ResNet-CPDS, respectively. These models differ in their encoder architecture. In our analysis, we explored skip connections to capture fine-grained details and contextual information. We got the best result by training the ResNet-CPDS model with accuracy and IoU of 98.75% and 72.04%, respectively. This model has three parts: encoder, bridge, and decoder. The input shape of the image is 256×256, which is downsampled to $256 \rightarrow 128 \rightarrow 64 \rightarrow 32 \rightarrow 16 \rightarrow 8$, and $8 \times 8 \times 2048$ acted as a bridge layer between the encoder and decoder part. In the decoder, we used the Conv2DTranspose for upsampling purposes. The upsampled output is then concatenated with the skip connection of the encoder, followed by convolution operation, batch normalization, and ReLU activation function twice. Although ResNet101V2 has around 44.7M parameters, the total parameters rose to 82.3M and 82.2M trainable parameters due to the modification in the pre-trained ResNet101V2 model. The significant rise in the number of parameters is due to using many filters in the decoder arm. The output shape in the decoder arm is as follows: $(16 \times 16 \times 1024)$, $(32 \times 32 \times 512)$, $(64 \times 64 \times 256)$, $(128 \times 128 \times 128)$, $(256 \times 256 \times 128)$. The output of $(256 \times 256 \times 128)$ is then passed to the convolutional layer with the sigmoid activation function, giving the final output masked image.

Figures 4, 5, and 6 depict the accuracy and loss curves for the following pairs of models: (i) ResNet101 and EfficientNetB3, (ii) ResNet101 and EfficientNetB6, and (iii) ResNet152V2 and EfficientNetB7, respectively. The first column of these figures illustrates the epoch versus training/validation accuracy curve. In contrast, the second column displays the epoch versus training/validation loss curve. Figures 7, 8, and 9 showcase the accuracy and loss curves for the following models: (i) ResNet50 with freezing, (ii) ResNet50 without freezing, and (iii) EfficientNetB7. Subsequently, Fig. 10 presents the epoch versus accuracy and epoch versus loss curves for the proposed ResNet-CPDS model.

Table 1 presents a summary of the performance of various models usable for CPDS. It is found that ResNet50 without freezing achieved the highest scores in terms of IoU, accuracy, precision, and recall among all other models. On the other

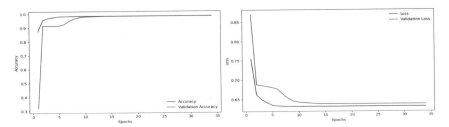

Fig. 4 Accuracy and loss graphs for ResNet101 and EfficientNetB3

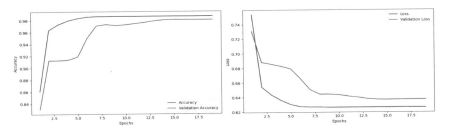

Fig. 5 Accuracy and loss graphs for ResNet101 and EfficientNetB6

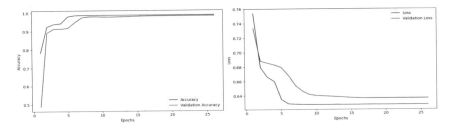

Fig. 6 Accuracy and loss graphs for ResNet152V2 and EfficientNetB7

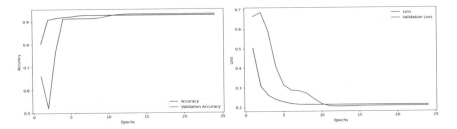

Fig. 7 Accuracy and loss graphs for ResNet50 with freezing

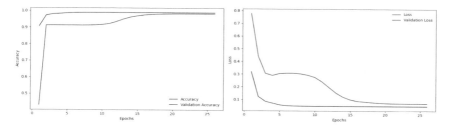

Fig. 8 Accuracy and loss graphs for ResNet50 without freezing

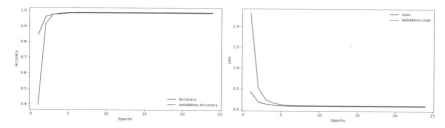

Fig. 9 Accuracy and loss graphs for EfficientNetB7

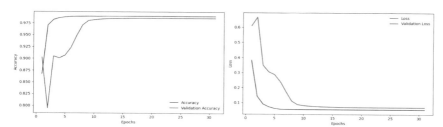

Fig. 10 Accuracy and loss graphs for ResNet-CPDS

Table 1 Performance comparison for accuracy, IoU, recall, and precision

Model	Accuracy (%)	IoU (%)	Recall (%)	Precision (%)
ResNet101 and EfficientNetB3	98.17	14.81	86.19	96.41
ResNet101 and EfficientNetB6	98.66	15.55	92.13	96.28
ResNet152V2 and EfficientNetB7	98.65	15.29	92.16	96.03
ResNet50 with freezing	92.65	20.48	38.13	72.64
ResNet50 without freezing	98.91	66.01	92.91	97.75
EfficientNetB7	**98.97**	**62.32**	**93.22**	**98.07**
ResNet-CPDS	**98.75**	**72.04**	**91.53**	**97.52**

Table 2 Performance comparison for training and validation loss

Model	Training loss	Validation loss
ResNet101 and EfficientNetB3	0.6313	0.6376
ResNet101 and EfficientNetB6	0.6258	0.6366
ResNet152V2 and EfficientNetB7	0.6275	0.6363
ResNet50 with freezing	0.2133	0.2072
ResNet50 without freezing	0.0486	0.0700
EfficientNetB7	**0.0559**	**0.0704**
ResNet-CPDS	**0.0775**	**0.0912**

hand, ResNet101 and EfficientNetB3 demonstrated the lowest accuracy and IoU, while ResNet50 with freezing exhibited the lowest recall and precision. Remarkably, the proposed ResNet-CPDS (modified ResNet101V2) model displayed significant improvements of 6.03% in IoU, 0.06% in accuracy, 0.31% in precision, and 0.32% in recall compared to ResNet50 without freezing model. Additionally, EfficientNetB7 without freezing, and ResNet-CPDS showed notable enhancements of 0.8% in accuracy, 57.23% in IoU, 55.09% in recall, and 25.43% in precision compared to the worst-performing models (a) ResNet101 and EfficientNetB3 and (b) ResNet50 with freezing. Table 2 compared all the models concerning the training and validation loss while achieving the highest accuracy and IoU values. It is evident that Efficient-NetB7 without freezing and ResNet-CPDS models have exhibited nearly optimal loss values alongside the best IoU, accuracy, recall, and precision scores. This observation reflects the stability and robustness of EfficientNetB7 without freezing and ResNet-CPDS models.

In the context of CPDS, the IoU is a crucial metric in assessing the effectiveness of any CNN model. As depicted in the above-presented results, the ResNet-CPDS model exhibits the highest IoU compared to other models. To further evaluate its performance, we computed the DSC, which yields an impressive score of 83.59%. These outcomes underscore the significance of adaptability, feature refinement, and skip connections in enhancing performance. Figures 11, 12, and 13 showcase positively detected outputs with higher IoU values generated by the ResNet-CPDS model. Conversely, Fig. 14 displays negatively detected outcomes with lower IoU values produced by the same model. For a more comprehensive performance analysis, we plotted the IoU values against the number of epochs required to train the ResNet-CPDS model in Fig. 15a. During the initial training phase, a notable difference exists between the IoU and validation IoU, but it gradually decreases, indicating the model's stability and convergence. Figure 15b illustrates the training and validation graph of the DSC versus epoch. It shows a gradual increase in DSC values with the number of epochs, highlighting the model's optimality. However, in deciding the suitability and practicality of a CPDS system, IoU holds the utmost importance. Based on the comprehensive statistical and visual results, it becomes evident that the ResNet-

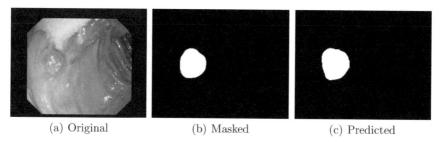

(a) Original   (b) Masked   (c) Predicted

Fig. 11 Positively detected regions by ResNet-CPDS model

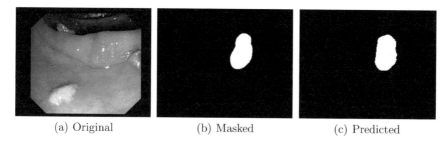

(a) Original   (b) Masked   (c) Predicted

Fig. 12 Positively detected regions by the ResNet-CPDS model

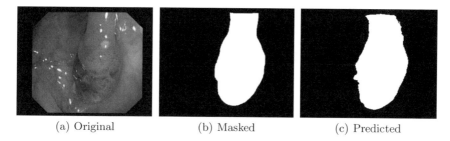

(a) Original   (b) Masked   (c) Predicted

Fig. 13 Positively detected regions by ResNet-CPDS model

CPDS model achieves better IoU, accompanied by nearly optimal accuracy, recall, precision, training loss, and validation loss. As a result, ResNet-CPDS emerges as a stable and reliable choice for the CPDS system.

5 Conclusions and Future Directions

This paper explored CNN-based architectures, focusing on the ResNet101V2 model, to develop an automated CPDS system. Among the various architectures, the modified ResNet101V2 U-Net is the most promising performer, exhibiting high accuracy, recall, precision, and IoU scores. To further enhance its performance, potential

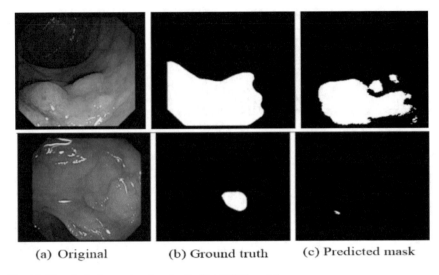

    (a) Original   (b) Ground truth   (c) Predicted mask

Fig. 14 Negatively detected regions by ResNet-CPDS model

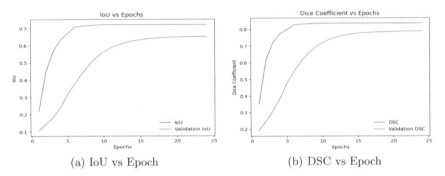

     (a) IoU vs Epoch   (b) DSC vs Epoch

Fig. 15 Results for the ResNet-CPDS model

strategies like increasing training epochs, employing data augmentation techniques, adjusting convolutional filters, and fine-tuning hyperparameters were identified. As a result, we proposed an improved system, ResNet-CPDS, for CPDS, and evaluated it using the CVC-ClinicDB dataset. The ResNet-CPDS model demonstrated comparable performance of 98.75% accuracy, and 72.04% IoU to state-of-the-art CPDS models despite the challenges posed by limited quality datasets, making it a viable choice for a cost-effective operational polyp segmentation system. However, there are ample opportunities for improvement, paving the way for future research in medical image segmentation techniques concerning the following points:

- Increase the dataset size to improve model performance and generalization.
- Explore transfer learning techniques to leverage pre-trained CPDS models.
- Develop a real-time CPDS system to integrate with the clinical workflows.

References

1. Morgan E, Arnold M, Gini A, Lorenzoni V, Cabasag CJ, Laversanne M, Vignat J, Ferlay J, Murphy N, Bray F (2023) Global burden of colorectal cancer in 2020 and 2040: incidence and mortality estimates from globocan. Cancers 72(2):338–344
2. Chatterjee R, Roy S, Islam SH (2021) Trident U-Net: an encoder fusion for improved biomedical image segmentation. In: Bioengineering and biomedical signal and image processing. Springer International Publishing, Cham, pp 141–154
3. Mo X, Tao K, Wang Q, Wang G (2018) An efficient approach for polyps detection in endoscopic videos based on faster R-CNN. In: 24th international conference on pattern recognition (ICPR 2018), pp 3929–3934. https://doi.org/10.1109/ICPR.2018.8545174
4. Jha D, Ali S, Tomar NK, Johansen HD, Johansen DD, Rittscher J, Riegler MA, Halvorsen P (2021) Real-time polyp detection, localization and segmentation in colonoscopy using deep learning. IEEE Access 9:40496–40510. https://doi.org/10.1109/ACCESS.2021.3063716
5. Shrestha S, Khanal B, Ali S (2020) Ensemble u-net model for efficient polyp segmentation. In: Proceedings of the MediaEval 2020 workshop
6. Bernal J, Sanchez F, Fernandez-Esparrach G, Gil D, Rodriguez de Miguel C, Vilarino F (2015) Wm-dova maps for accurate polyp highlighting in colonoscopy: validation versus saliency maps from physicians. Computerized Med Imaging Graph 43. https://doi.org/10.1016/j.compmedimag.2015.02.007
7. Ronneberger O, Fischer P, Brox T (2015) U-Net: convolutional networks for biomedical image segmentation
8. Zhou Z, Rahman Siddiquee MM, Tajbakhsh N, Liang J (2018) Unet++: a nested u-net architecture for medical image segmentation. In: Deep learning in medical image analysis and multimodal learning for clinical decision support. Springer International Publishing, Cham, pp 3–11
9. Huang H, Lin L, Tong R, Hu H, Zhang Q, Iwamoto Y, Han X, Chen YW, Wu J (2020) Unet 3+: a full-scale connected unet for medical image segmentation

Properties of m-Bonacci Words

Kalpana Mahalingam and Helda Princy Rajendran

Abstract The m-Bonacci word is the unique fixed point of the morphism φ_m : $0 \to 01$, $1 \to 02$, $2 \to 03, \ldots, (m-2) \to 0(m-1)$, $(m-1) \to 0$. The finite m-Bonacci word $w_{n,m}$ is defined as $w_{n,m} = \varphi_m^n(0)$. We study some combinatorial properties of finite m-Bonacci words. We find the values of n, such that $w_{n,m}$ is square free. We prove that $w_{n,m}$ is primitive and has a unique representation as a product of two palindromes. We also show that the language $W_m^0 = \{w_{n,m} : n \geq 0\}$ is context-free free and not dense.

Keywords m-Bonacci word · Primitive · Palindrome · Context-free free

1 Introduction

The sequence of Fibonacci words plays a very important role in the combinatorial theory of free monoids. Fibonacci words were first introduced by Knuth in [19]. Knuth defined the sequence of Fibonacci words as $f_1 = a$, $f_2 = b$ and for $n \geq 1$, $f_{n+2} = f_{n+1} \cdot f_n$, where "·" denotes the word/string catenation, a and b are two letters. The sequence of such Fibonacci words are denoted as $F_{a,b}$, where $F_{a,b} = \{a, b, ba, bab, babba, \ldots\}$. In 1980, Berstel gave an equivalent definition for the sequence of Fibonacci words using the morphism $\theta : a \to ab$, $b \to a$ [3]. The n-th Fibonacci word f_n is given by $f_n = \theta^{n-1}(a)$ for all $n \geq 1$.

The Fibonacci word has many generalizations. To name a few, indexed Fibonacci word [5], mapped shuffled Fibonacci languages [13], k-Fibonacci words [22], (n, i)-Fibonacci words [23, 24], involutive Fibonacci words [17], Fibonacci sequence on an infinite alphabet [27], m-Bonacci words [4], k-Bonacci words on infinite alphabet [8].

K. Mahalingam · H. P. Rajendran (✉)
Department of Mathematics, Indian Institute of Technology Madras, 600036 Chennai, India
e-mail: heldaprincy96@gmail.com

K. Mahalingam
e-mail: kmahalingam@iitm.ac.in

© The Author(s), under exclusive license to Springer Nature Singapore Pte Ltd. 2024
D. Giri et al. (eds.), *Proceedings of the Tenth International Conference on Mathematics and Computing*, Lecture Notes in Networks and Systems 963,
https://doi.org/10.1007/978-981-97-2069-9_9

The m-Bonacci word is a natural generalization of the Fibonacci word. The infinte m-Bonacci word is the unique fixed point of the following morphism.

$$\varphi_m : 0 \to 01, \ 1 \to 02, \ 2 \to 03, \ldots, (m-2) \to 0(m-1), \ (m-1) \to 0$$

The m-Bonacci word belongs to the class of Arnoux-Rauzy words and hence, episturmian word. An Arnoux-Rauzy word is an infinite word w, such that for any non-negative integer n, w has exactly one right-special factor and exactly one left-special factor of length n [1]. Various researchers have worked on the infinite m-Bonacci word. To name a few, Brinda et al. studied the balance property of m-Bonacci word ([4]), Jahannia et al. studied palindromic Ziv-lempel factorization, Crochemore factorization and closed Ziv-lempel factorization of m-Bonacci word [14, 15], Bo Tan et al. studied the Lyndon factorization and the singular factorization of Tribonacci word [25]. For other properties, refer [10].

In this paper, we discuss some combinatorial properties of the finite m-Bonacci words such as primitivity, palindromic factors, and conjugates. The finite m-Bonacci word $w_{n,m}$ is defined as $w_{n,m} = \varphi_m^n(0)$. This work is a preliminary work, as we have numerous things to explore for the infinite m-Bonacci word. We show that the finite m-Bonacci word is primitive and it can be uniquely written as a product of two palindromes. We find the values of n, such that $w_{n,m}$ is square free. We define the standard m-Bonacci language as the set of all finite m-Bonacci words. We show that the standard m-Bonacci language is not dense and context-free free.

The paper is organized as follows. In Sect. 2, we give some basic definitions and notations. In Sect. 3, we give some properties of finite m-Bonacci words. In Sect. 4, we discuss the properties of the standard m-Bonacci language $W_m^0 = \{w_{n,m} : n \geq 0\}$. We end the paper with a few concluding remarks.

2 Preliminaries

We denote the alphabet set by Σ. A word w over Σ is a sequence of elements of Σ. A word can be finite or infinite. The set of all finite words over Σ is denoted as $\Sigma^* = \Sigma^+ \cup \{\lambda\}$, where Σ^+ is the set of all non-empty words and λ denotes the empty word. A *morphism* on an alphabet Σ is a mapping $\phi : \Sigma^* \to \Sigma^*$ which satisfies $\phi(xy) = \phi(x)\phi(y)$ for every $x, y \in \Sigma^*$.

The length of a word $u \in \Sigma^*$ is the number of letters in u, and is denoted by $|u|$. For any $a \in \Sigma$, $|u|_a$ denote the number of occurrences of the letter a in the word u. Let $\Sigma = \{0, 1, 2, \ldots, m-1\}$ and let $w \in \Sigma^*$, we denote the vector $(|w|_0, |w|_1, \ldots, |w|_{m-1})$ by $L(w)$. A word v is a factor of u, if there exist $x, y \in \Sigma^*$, such that $u = xvy$. A word v is said to be a prefix (respectively, suffix) of u, if there exists $x \in \Sigma^*$ such that $u = vx$ (respectively, $u = xv$). Let $u = u_1u_2u_3 \ldots u_n \in \Sigma^*$, where $u_i \in \Sigma$. The reverse of u is denoted by u^R is defined as $u^R = u_nu_{n-1} \ldots u_1$. A word u is a palindrome if $u = u^R$. Let P denote the set of all palindromes over the alphabet Σ.

A word $w \in \Sigma^*$ is called a conjugate of $u \in \Sigma^*$ if there exist x, $y \in \Sigma^*$ such that $u = xy$ and $w = yx$. We denote the set of all conjugates of a word $u \in \Sigma^*$ by $C(u)$. Let $u = u_1 u_2 u_3 \ldots u_n, u_i \in \Sigma, 1 \le i \le n$. We say that an integer k to be a period of u if $i \equiv j \pmod{k}$ implies $u_i = u_j$. For $u = xy, uy^{-1} = x$, where y^{-1} is called as inverse of y. The inverse of any word $u = u_1 u_2 u_3 \ldots u_n$ ($u_i \in \Sigma$), denoted as u^{-1}, is defined as $u^{-1} = u_n^{-1} u_{n-1}^{-1} \ldots u_1^{-1}$ [25].

We recall the definition of infinite m-Bonacci word as follows.

Definition 1 The *infinite m-Bonacci word* is the unique fixed point of the morphism

$$\varphi_m : 0 \to 01, \ 1 \to 02, \ 2 \to 03, \ldots, (m-2) \to 0(m-1), \ (m-1) \to 0$$

We recall the definition of finite m-Bonacci word as follows. In [8], Gharegani et al. generalized the following definition to infinite alphabets and defined m-Bonacci words over an infinite alphabet.

Definition 2 For each $n \ge 0$, finite m-Bonacci word is defined as

$$w_{n.m} = \varphi_m^n(0)$$

Equivalently, finite m-Bonacci word can be defined as

$$w_{n,m} = \begin{cases} 0 & \text{if } n = 0 \\ w_{n-1,m} w_{n-2,m} \ldots w_{0,m} n & \text{if } 1 \le n < m \\ w_{n-1,m} w_{n-2,m} \ldots w_{n-m,m} & \text{if } n \ge m \end{cases}$$

In [2], Ammar et al. showed that the above two definitions for the finite m-Bonacci word are equivalent. For $m = 2$, it is nothing but the standard Fibonacci word. For example, when $m = 4$, the sequence of finite 4-Bonacci word is given below.

$$0, 01, 0102, 01020103, 010201030102010, \ldots$$

Next, we recall the definition of m-Bonacci sequence.

Definition 3 [16] The m-Bonacci sequence $\{Z_{n,m}\}_{n \ge 1}$ is defined by

$$Z_{i,m} = 0, \ -(m-2) \le i \le 0, \ Z_{1,m} = 1$$

and for $n \ge 2$,

$$Z_{n,m} = \sum_{i=n-m}^{n-1} Z_{i,m}$$

Each $Z_{i,m}$ is called an *m-Bonacci number*.

For example, when $m = 5$, the sequence is $0, 0, 0, 0, 1, 1, 2, 4, 8, 16, 31 \ldots$.

We recall the following results from [6, 20, 21].

Lemma 1 *[20] Let u, $v \in \sum^{+}$. Then $uv = vu$ implies that u and v are powers of a common word.*

Palindromes are also known as symmetric words. In 1993, Wai-fong Chuan proved the following.

Lemma 2 *[6] If a word has more than one representation as a product of two symmetric words, then it is a power of another word.*

Lemma 3 *[21] The cardinality of the class of conjugates of a primitive word w of length n is n, i.e., $|c(w)| = n$.*

3   Properties of m-Bonacci Word

In this section, we study some combinatorial properties of finite m-Bonacci words. In [11], the author showed that the length of Fibonacci word $w_{n,2}$ is $n + 2$-th Fibonacci number. In [18], the author showed that the length of the finite m-Bonacci word is $Z_{n+2,m}$.

Remark 1 [18] Let $m \geq 2$. Then $|w_{n,m}| = Z_{n+2,m}$

We know that $|w_{n,m}| = |w_{n,m}|_0 + |w_{n,m}|_1 + \ldots + |w_{n,m}|_{(m-1)}$. For the Fibonacci word, we have $|w_{n,2}|_0 = Z_{n+1,2}$ and $|w_{n,2}|_1 = Z_{n,2}$ [11]. The same result can be generalized to any $m \geq 3$. In the following result, we calculate the number of occurrences of each alphabet in the n-th finite m-Bonacci word.

Proposition 1 *Let $n \geq 0$ and $m \geq 2$. Then*

$$L(w_{n,m}) = (Z_{n+1,m}, Z_{n,m}, Z_{n-1,m}, \ldots, Z_{n-m+2,m})$$

Proof. The proof follows by induction on n.   □

In the following result, we show that the word $w_{n-2,m}^2$ is a prefix of the n-th finite m-Bonacci word $w_{n,m}$, $n \geq m + 2$. We omit the proof, as it can be easily verified by direct calculation using the definition of the finite m-Bonacci word.

Proposition 2 *For $n \geq m + 2$, $w_{n,m} = w_{n-2,m}^2 y_{n,m}$, where*

$$y_{n,m} = \prod_{i=m+3}^{2m+1} w_{n-i,m} \prod_{i=m+2}^{2m} w_{n-i,m} \cdots \prod_{i=4}^{m+2} w_{n-i,m} \prod_{i=3}^{m} w_{n-i,m}.$$

□

In the following result, we show that, for $m \geq 3$, $w_{m,m}^3$ is a factor of the n-th finite m-Bonacci word $w_{n,m}$ for all $n \geq 2m + 1$ and for $m = 2$, the word $w_{2,2}^3$ is a factor of $w_{n,2}$, $n \geq 6$.

Proposition 3 *Let* $m \geq 3$. *For* $n \geq 2m + 1$, *the word* $w_{m,m}^3$ *is a factor of* $w_{n,m}$. *Also,* $w_{m,m}^3$ *is not a factor of* $w_{n,m}$, $n < 2m + 1$. *For* $m = 2$ *and* $n \geq 6$, $w_{2,2}^3$ *is a factor of* $w_{n,2}$. *For* $n < 6$, $w_{2,2}^3$ *is not a factor of* $w_{n,2}$.

Proof. Let $m \geq 3$. Since $w_{n,m}$ is a prefix of $w_{n+1,m}$, it is enough to show that $w_{m,m}^3$ is a factor of w_{2m+1}. By Proposition 2, $w_{m,m}^2$ is a prefix of $w_{m+2,m}$. Since $m \geq 3$, $2m - 1 \geq m + 2$ and hence, $w_{m,m}^2$ is a prefix of w_{2m-1}. By the definition, we have, $w_{2m,m} = w_{2m-1,m} w_{2m-2,m} \cdots w_{m,m}$. Thus, $w_{m,m}$ is a suffix of $w_{2m,m}$. Also, $w_{2m+1,m} = w_{2m,m} w_{2m-1,m} \cdots w_{m+1,m}$. Hence, $w_{m,m}^3$ is a factor of $w_{2m+1,m}$.

By direct verification, it is easy to verify that $w_{m,m}^3$ is not a factor of $w_{m+2,m}$ and hence, $w_{m,m}^3$ is not a factor of $w_{n,m}$, $n \leq m + 2$. Now, the only thing left to prove is $w_{m,m}^3$ is not a factor of $w_{n,m}$, $m + 2 < n < 2m + 1$. If not, choose l such that l is the least positive integer and $w_{m,m}^3$ is a factor of $w_{l,m}$. By definition, $w_{l,m} = w_{l-1,m} w_{l-2,m} \cdots w_{l-m,m}$. Since l is the least positive integer, $w_{m,m}^3$ is not a factor of both $w_{l-1,m}$ and $w_{l-2,m} w_{l-3,m} \cdots w_{l-m,m}$. Hence, for some $u, v \in \Sigma^+$, we have the following.

$$w_{m,m}^3 = uv, \text{ where } w_{l-1,m} = xu \text{ and } w_{l-2,m} w_{l-3,m} \cdots w_{l-m,m} = vy \qquad (1)$$

Since $m + 2 < l < 2m + 1$, $w_{m,m}$ is not a suffix of $w_{l-1,m}$ and $w_{m,m}$ is a prefix of $w_{l-2,m}$. This gives a contradiction to Eq. 1. Hence, $w_{m,m}^3$ is not a factor of $w_{l,m}$.

For $m = 2$, the result can be verified directly.   □

From Proposition 3, we get that $w_{n,m}$ is not cube free for $n \geq 2m + 1$. In [9], the author showed that the infinite m-Bonacci word is fourth power free. For $n = m + 1$, $w_{m+1,m} = w_{m,m} w_{m-1,m} \cdots w_{1,m}$. We know that, 0 is a prefix for any $w_{n,m}$. Also, from Proposition 5, it is clear that $w_{m,m}$ is a palindrome. Since 0 is a suffix of $w_{m,m}$ and 0 is a prefix of $w_{m-1,m}$, 00 is a factor of $w_{m+1,m}$. Hence, $w_{m+1,m}$ is not square free. In the following proposition, we show that the initial $m + 1$ finite m-Bonacci words are square free.

Proposition 4 *Let* $m \geq 2$. *The finite* m-*Bonacci word* $w_{n,m}$ *is square free for* $0 \leq n \leq m$.

Proof. We have $w_{0,m} = 0$ and $w_{1,m} = 01$. For $1 \leq n \leq (m - 1)$, by the definition of finite m-Bonacci word, we have, $w_{n,m} = w_{n-1,m} w_{n-2,m} \cdots w_{0,m} n$. This shows that $w_{n,m}$ is square free for $0 \leq n \leq m$.   □

For $n \geq 1$, let $D_{n,m} = w_{n-1,m} w_{n-2,m} \cdots w_{0,m}$ and $D_{0,m} = \lambda$. One example is given below.

$$D_{1,4} = w_{0,4} = 0, \quad D_{2,4} = w_{1,4} w_{0,4} = 010, \quad D_{3,4} = w_{2,4} w_{1,4} w_{0,4} = 0102010,$$

$$D_{4,4} = w_{3,4} w_{2,4} w_{1,4} w_{0,4} = 01020100102010$$

In [25], authors showed that $D_{n,3}$ is a palindrome. For $m \geq 3$, Ammar et al. showed that $D_{n,m}$ is a palindrome for $n \geq 1$ [2]. In the following Lemma, we give a relation between $D_{n,m}$ and $D_{n+1,m}$.

Lemma 4 Let $m \geq 2$ and $n \geq 1$. Then, $\varphi_m(D_{n,m}) = D_{n+1,m}0^{-1}$.

Proof. By the definition of $D_{n,m}$, we have the following.

$$
\begin{aligned}
\varphi_m(D_{n,m}) &= \varphi_m(w_{n-1,m} w_{n-2,m} \ldots w_{0,m}) \\
&= w_{n,m} w_{n-1,m} \ldots w_{1,m} = D_{n+1,m}0^{-1} \quad\quad (2)
\end{aligned}
$$

Hence, the result.   □

In the following Lemma, we show that the m-th finite m-Bonacci word $w_{m,m}$ can be written as $D_{m-i,m} \prod\limits_{j=m-i}^{m-1} j D_{j,m}$ for $1 \leq i \leq m-1$. The proof of the Lemma is easy to observe from the definition of $w_{n,m}$ and $D_{n,m}$.

Lemma 5 Let $m \geq 2$ and $1 \leq i \leq m-1$. Then, $w_{m,m} = D_{m-i,m} \prod\limits_{j=m-i}^{m-1} j D_{j,m}$. For $m \geq 2$, $n \geq m$ and $1 \leq i \leq m-1$, we define the following.   □

$$
P^i_{n,m} = w_{n-1,m} w_{n-2,m} \ldots w_{n-i,m}, \quad S^i_{n,m} = w_{n-(i+1),m} w_{n-(i+2),m} \ldots w_{n-m,m}
$$

Hence, $w_{n,m} = P^i_{n,m} S^i_{n,m}$. Now, we define the i-th m-Bonacci shift of the finite m-Bonacci word $w_{n,m}$ as follows.

Definition 4 Let $m \geq 2$, $n \geq m$ and $1 \leq i \leq m-1$. The i-th m-Bonacci shift of the finite m-Bonacci word $w_{n,m}$, denoted as $\sigma^i_m(w_{n,m})$, is defined as $\sigma^i_m(w_{n,m}) = S^i_{n,m} P^i_{n,m}$.

In [2], Ammar et. al. showed that, for $n \geq m$, $1 \leq i \leq m-1$, the longest common prefix of $w_{n,m}$ and $\sigma^i_m(w_{n,m})$ is $D_{n-i,m}$. Now, we define $E^i_{n,m}$ and $F^i_{n,m}$ for $n \geq m$ and $1 \leq i \leq m-1$ as follows. Since $D_{n-i,m}$ is the longest common prefix of $w_{n,m}$ and $\sigma^i_m(w_{n,m})$, we define the remaining suffix of $w_{n,m}$ and $\sigma^i_m(w_{n,m})$ as $E^i_{n,m}$ and $F^i_{n,m}$, respectively. This is shown below.

$$
w_{n,m} = D_{n-i,m} E^i_{n,m}; \quad \sigma^i_m(w_{n,m}) = D_{n-i,m} F^i_{n,m}
$$

From Lemma 5, we have $E^i_{m,m} = \prod\limits_{j=m-i}^{m-1} j D_{j,m}$. We illustrate $P^i_{n,m}$, $S^i_{n,m}$, $E^i_{n,m}$, $F^i_{n,m}$ and $\sigma^i_m(w_{n,m})$ with the following example.

Example 1 The 4-th 4-Bonacci word is $w_{4,4} = 010201030102010$.

$P^1_{4,4}$	$S^1_{4,4}$	$\sigma^1_4(w_{4,4})$	$E^1_{4,4}$	$F^1_{4,4}$
01020103	0102010	(0102010)(01020103)	30102010	01020103

$P^2_{4,4}$	$S^2_{4,4}$	$\sigma^2_4(w_{4,4})$	$E^2_{4,4}$	$F^2_{4,4}$
010201030102	010	(010)(010201030102)	201030102010	010201030102

$P^3_{4,4}$	$S^3_{4,4}$	$\sigma^3_4(w_{4,4})$	$E^3_{4,4}$	$F^3_{4,4}$
01020103010201	0	(0)(01020103010201)	10201030102010	01020103010201

In the following Lemma, we give a relation between $E^i_{n,m}$ and $E^i_{n-1,m}$.

Lemma 6 *Let $n \geq m+1$ and $1 \leq i \leq m-1$. Then, $E_{n,m} = 0^{-1}\varphi_m(E^i_{n-1,m})$.*

Proof. By the definition of $E^i_{n,m}$, we have $w_{n-1,m} = D_{n-1-i,m} E^i_{n-1,m}$. Now, we have the following.

$$w_{n,m} = \varphi_m(w_{n-1,m}) = \varphi_m(D_{n-1-i,m} E^i_{n-1,m})$$
$$= D_{n-i,m} 0^{-1}\varphi_m(E^i_{n-1,m}) \text{ (by Lemma 4)} \tag{3}$$

Hence, the result. □

From Lemma 5 and 6, we have the following. The following corollary gives a recurrence relation for $E^i_{n,m}$, $n \geq m$, $1 \leq i \leq m-1$.

Corollary 1 *Let $m \geq 2$ and $n \geq m$. For $1 \leq i \leq m-1$,*

$$E^i_{m,m} = \prod_{j=m-i}^{m-1} j D_{j,m}, \quad E^i_{n,m} = 0^{-1}\varphi_m(E^i_{n-1,m}), \quad n \geq m+1$$

□

From Example 1, one can verify that $E^i_{4,4}$ is the mirror image of $F^i_{4,4}$. This holds true for any n and m. For $m = 2, 3$, the result is already shown in the literature [7, 25]. Now, we prove for any $m \geq 2$. For any $w \in \Sigma^*$, we have $\varphi_m(w^R) = 0(\varphi_m(w))^R 0^{-1}$ [2]. In the following theorem, we show that the suffix $E^i_{n,m}$ of the n-th finite m-Bonacci word is the mirror image of the suffix of the i-th m-Bonacci shift of the n-th finite m-Bonacci word.

Theorem 1 *Let $m \geq 2$ and $n \geq m$. Then,*

$$\sigma^i_m(w_{n,m}) = D_{n-i,m}(E^i_{n,m})^R,$$

where $w_{n,m} = D_{n-i,m} E^i_{n,m}$, $1 \leq i \leq m-1$.

Proof. We prove this result by induction on n. From Lemma 5, we have $w_{m,m} = D_{m-i,m} \prod_{j=m-i}^{m-1} j D_{j,m}$. Also, we have the following

$$\sigma_m^i(w_{m,m}) = S_{m,m}^i P_{m,m}^i = w_{m-(i+1),m} \cdots w_{0,m} w_{m-1,m} w_{m-2,m} \cdots w_{m-i,m}$$
$$= D_{m-i,m} w_{m-1,m} w_{m-2,m} \cdots w_{m-1,m}$$

Now, it is enough to show that $\left(\prod_{j=m-i}^{m-1} j D_{j,m} \right)^R = \prod_{j=1}^{i} w_{m-j,m}$

$$\left(\prod_{j=m-i}^{m-1} j D_{j,m} \right)^R = \prod_{j=m-i}^{m-1} D_{j,m}^R j = \prod_{j=m-1}^{m-i} D_{j,m} j \text{ (since } D_{j,m} \text{ is a palindrome)}$$
$$= \prod_{j=1}^{i} w_{m-j,m} \text{ (by the definition of } D_{j,m} \text{ and } w_{n,m}, n < m) \text{ (4)}$$

This proves the induction case. Let $n \geq m + 1$. Assume that the result is true for all $k < n$. From Lemma 4 and 6, we have $\varphi_m(D_{n,m}) = D_{n+1,m}0^{-1}$ and $E_{n,m}^i = 0^{-1}\varphi_m(E_{n-1,m}^i)$. By the definition of $w_{n,m}$, we have the following.

$$\sigma_m^i(w_{n,m}) = S_{n,m}^i P_{n,m}^i = \prod_{j=i+1}^{m} w_{n-j,m} \cdot \prod_{j=1}^{i} w_{n-j,m}$$
$$= \prod_{j=i+1}^{m} \varphi_m(w_{n-1-j,m}) \cdot \prod_{j=1}^{i} \varphi_m(w_{n-1-j,m})$$
$$= \varphi_m \left(\prod_{j=i+1}^{m} w_{n-1-j,m} \cdot \prod_{j=1}^{i} w_{n-1-j,m} \right)$$
$$= \varphi_m(D_{n-1-i,m}(E_{n-1,m}^i)^R) \text{ (by induction hypothesis)}$$
$$= D_{n-i,m}0^{-1}\varphi_m((E_{n-1,m}^i)^R) \text{ (by Lemma 4)}$$
$$= D_{n-i,m}0^{-1}0(\varphi_m(E_{n-1,m}^i))^R 0^{-1} \text{ (5)}$$
$$\text{(since } \varphi_m(u^R) = 0(\varphi_m(u))^R 0^{-1} \text{for any } u)$$

Since $E_{n,m}^i = 0^{-1}\varphi_m(E_{n-1,m}^i)$, Eq. 5 concludes the induction case. Hence, the result. □

In [7], Aldo de Luca proved that the finite Fibonacci words are primitive. The same result can be extended to any finite m-Bonacci word. In the following proposition, we prove that the finite m-Bonacci words are primitive.

Theorem 2 *Let $m \geq 2$ and $n \geq 0$. The finite m-Bonacci word $w_{n,m}$ is primitive.*

Proof. We prove this by induction on n. From the definition of finite m-Bonacci word, we know that $w_{0,m} = 0$ and for $1 \leq n < m$, $w_{n,m} = w_{n-1,m} w_{n-2,m} \ldots w_{0,m} n$. Since the alphabet n appears exactly once in $w_{n,m}$, $w_{n,m}$ is primitive for $1 \leq n \leq m - 1$. Also, we have, $w_{m,m} = w_{m-1,m} w_{m-2,m} \ldots w_{0,m}$. The alphabet $(m - 1)$ occurs exactly once in the word $w_{m,m}$ and hence, $w_{m,m}$ is primitive. Let $n \geq m + 1$. Assume that, $w_{k,m}$ is primitive for every $k < n$. Now, we prove for $w_{n,m}$. Suppose $w_{n,m}$ is not primitive, then $w_{n,m} = u^t$ for some $u \in \Sigma^*$ and $t \geq 2$. From Theorem 1, we have $w_{n,m} = D_{n-1,m} E_{n,m}^1$, $w_{n-2,m} w_{n-3,m} \ldots w_{n-m,m} w_{n-1,m} = D_{n-1,m} (E_{n,m}^1)^R$. Since $w_{n,m} = u^t$, $D_{n-1,m} = u^{t_1} u_1$ and $E_{n,m}^1 = u_2 u^{t_2}$, where $u = u_1 u_2$ and $t = t_1 + t_2 + 1$. Since $D_{n-1,m}$ is a palindrome, we have

$$D_{n-1,m} = (u_1 u_2)^{t_1} u_1 = ((u_1 u_2)^{t_1} u_1)^R = (u_1^R u_2^R)^{t_1} u_1^R \tag{6}$$

Thus, $u_1 = u_1^R$ and $u_2 = u_2^R$. This implies, $(E_{n,m}^1)^R = (u_2(u_1 u_2)^{t_2})^R = E_{n,m}$, which is a contradiction. Hence, the result. $\qquad\square$

The following corollary follows directly from Lemma 3, Remark 1, and Theorem 2.

Corollary 2 *Let $n \geq 0$ and $m \geq 2$. Then $|C(w_{n,m})| = |C(w'_{n,m})| = Z_{n+2,m}$.*

$\qquad\square$

In [7], Aldo de Luca proved that any finite Fibonacci word has a unique representation as a product of two palindromes. Similar result holds for any m. First, we show that any finite m-Bonacci word $w_{n,m}$, $n \geq m$, can be written as a product of two palindromes.

Proposition 5 *Let $m \geq 2$. For $n \geq m$, the finite m-Bonacci word $w_{n,m}$ can be factorized into $u_{n,m} v_{n,m}$, where $u_{n,m}$ and $v_{n,m}$ are palindromes. Also,*

$$|u_{n,m}| = \begin{cases} 0 & \text{if } n = m \\ \displaystyle\sum_{i=2}^{n-m+1} Z_{i,m} & \text{otherwise} \end{cases}$$

Proof. We prove this result by induction on n.

Base case: When $n = m$, $w_{m,m} = w_{m-1,m} w_{m-2,m} \ldots w_{0,m}$. Since $w_{m,m} = D_{m,m}$, $w_{m,m}$ is a palindrome. Therefore, $w_{m,m} = \lambda w_{m,m}$. This proves the base case. Let $n \geq m + 1$. Assume that, the result is true for any $t < n$.

Induction case: We have, $w_{n,m} = w_{n-1,m} w_{n-2,m} \ldots w_{n-m,m}$. Therefore, $w_{n,m} w_{n-m-1,m} = w_{n-1,m} w_{n-1,m}$. By induction, $w_{n-1,m} = u_{n-1,m} v_{n-1,m}$, where $u_{n-1,m}$ and $v_{n-1,m}$ are palindromes. Hence,

$$w_{n,m} w_{n-m-1,m} = u_{n-1,m} v_{n-1,m} u_{n-1,m} v_{n-1,m} \tag{7}$$

From Eq. 7, we have $w_{n-m-1,m}$ is a suffix of $v_{n-1,m}$. Let $v_{n-1,m} = x w_{n-m-1,m}$ for some $x \in \Sigma^*$. Since $v_{n-1,m}$ is a palindrome,

$$v_{n-1,m} = x w_{n-m-1,m} = w^R_{n-m-1,m} x^R \tag{8}$$

Using Eqs. 7 and 8, we get the following.

$$w_{n,m} = u_{n-1,m} w^R_{n-m-1,m} x^R u_{n-1,m} x \tag{9}$$

From Eq. 9, we get that $w_{n-1,m} = u_{n-2,m} w^R_{n-m-2,m} y^R u_{n-2,m} y$, for some $y \in \Sigma^*$. But, by induction, we have, $w_{n-1,m} = u_{n-1,m} v_{n-1,m}$ and $|u_{n-1,m}| = \sum_{i=2}^{n-m} Z_{i,m}$ and $|u_{n-2,m}| = \sum_{i=2}^{n-m-1} Z_{i,m}$. Since $|w_{n-m-2}| = Z_{n-m,m}$, $u_{n-1,m} = u_{n-2,m} w^R_{n-m-2,m}$. Hence, we have the following.

$$
\begin{aligned}
u_{n-1,m} w^R_{n-m-1,m} &= u_{n-2,m} w^R_{n-m-2,m} w^R_{n-m-1,m} \\
&= u_{n-3,m} w^R_{n-m-3,m} w^R_{n-m-2,m} w^R_{n-m-1,m} \\
&\;\;\vdots \\
&= w^R_{0,m} w^R_{1,m} \cdots w^R_{n-m-1,m} \\
&= (w_{n-m-1,m} w_{n-m-2,m} \cdots w_{0,m})^R = D_{n-m,m} \tag{10}
\end{aligned}
$$

Since $D_{n-m,m}$ is a palindrome, $u_{n-1,m} w^R_{n-m-1,m}$ is a palindrome. Let $u_{n,m} = u_{n-1,m} w^R_{n-m-1,m}$ and $v_{n,m} = x^R u_{n-1,m} x$. Since $u_{n-1,m}$ is a palindrome, $v_{n,m}$ is a palindrome. Also we have, $|u_{n,m}| = |u_{n-1,m}| + |w_{n-m-1,m}|$. By induction, $|u_{n-1,m}| = \sum_{i=2}^{n-m} Z_{i,m}$. From Remark 1, $|w_{n-m-1,m}| = Z_{n-m+1,m}$. This proves the induction case. Hence, the result.   □

In the following theorem, we prove that any finite m-Bonacci word has a unique representation as a product of two palindromes.

Theorem 3 *Every finite m-Bonacci word has a unique representation as a product of two palindromes*

Proof. *Existence*: For $n \geq m$, from Proposition 5, we have that $w_{n,m} = u_{n,m} v_{n,m}$, where $u_{n,m}$ and $v_{n,m}$ are palindromes. Now, for $n \leq m-1$, from Definition 2, it is easy to observe that $w_{n,m} = D_{n,m} n$ for $n < m$. Since $D_{n,m}$ is a palindrome, $w_{n,m}$ can be written in the form $u_{n,m} v_{n,m}$, where $u_{n,m} = D_{n,m}$ and $v_{n,m} = n$ and $u_{n,m}$ and $v_{n,m}$ are palindromes, for $0 \leq n \leq m-1$.

Uniqueness: Assume that, $w_{n,m}$ has another representation as a product of two palindromes. By Lemma 2, $w_{n,m}$ is a power of another word. This gives a contradiction to the fact that $w_{n,m}$ is primitive. Hence, the result.   □

4   Properties of the Standard m-Bonacci Language

Let $W_m^0 = \{w_{n,m}\}_{n=0}^{\infty}$. We call W_m^0 as the standard m-Bonacci language. In this section, we discuss the properties of the standard m-Bonacci language. In [26], Yu et al. showed that $F_{a,b} \cap P = \{1, 0, 010\}$, where $F_{a,b} = \{w_n, n \geq 1 : w_1 = 1, w_2 = 0, w_n = w_{n-2}w_{n-1}\}$. From this, we get that $W_2^0 \cap P = \{0, 010\}$. In the following result, we show that the number of palindromes in the standard m-Bonacci language is exactly two. We omit the proof, as it is similar to the proof of $F_{a,b} \cap P = \{1, 0, 010\}$ [26].

Proposition 6   *Let $m \geq 2$. Then, $W_m^0 \cap P = \{0, w_{m,m}\}$.*   □

A language L is dense if for any $u \in \Sigma^*$, $\Sigma^* u \Sigma^* \cap L \neq \emptyset$. In [26], Yu et al. showed that W_2^0 is not dense. This result can be generalized for any $m \geq 0$. That is, W_m^0 is not dense.

Proposition 7   *Let $m \geq 2$. W_m^0 is not dense.*

Proof.   Let $u = 0000 \in \Sigma^*$. Since every finite m-Bonacci word is fourth power free, $\Sigma^* u \Sigma^* \cap W_m^0 = \emptyset$. Hence, the result.   □

A language L is called context-free free if every infinite subset of L is not context-free. If an infinite language L is a context-free language, then there exist $x_1, x_2, x_3, x_4, x_5 \in \Sigma^*$ and $|x_2 x_4| > 0$, $\{x_1 x_2^n x_3 x_4^n x_5 : n \geq 0\} \subseteq L$ [12]. In [26], Yu et. al. showed that W_2^0 is context-free free. The same result holds for any $m \geq 2$. In the following result, we show that the standard m-Bonacci language is context-free free.

Proposition 8   *Let $m \geq 2$. W_m^0 is context-free free.*

Proof.   Suppose W_m^0 is not context-free free, then there exist $x_1, x_2, x_3, x_4, x_5 \in \Sigma^*$, such that $|x_2 x_4| \geq 1$ and $\{x_1 x_2^n x_3 x_4^n x_5 : n \geq 0\} \subseteq W_m^0$. Since $x_1 x_2^n x_3 x_4^n x_5 \in W_m^0$, there exists $k \geq 0$, $x_1 x_2^n x_3 x_4^n x_5 = w_{k,m}$. By the definition of finite m-Bonacci word, we have $|w_{n,m}| < |w_{n+1,m}|$ for any $n \geq 0$. Now, by the definition of $w_{k,m}$, we have, $x_1 x_2^n x_3 x_4^n x_5 = w_{k,m} = w_{k-1,m} w_{k-2,m} \cdots w_{k-m,m}$. Then, $w_{k-1,m} w_{k-2,m} \cdots w_{k-m+1,m} = u$, where $x_1 x_2^n x_3 x_4^n x_5 = uv$, $|u| > |v|$. Now, $w_{k+1,m} = w_{k,m} w_{k-1,m} w_{k-2,m} \cdots w_{k-m+1,m} = x_1 x_2^n x_3 x_4^n x_5 u$. If $x_1 x_2^n x_3 x_4^n x_5 u = x_1 x_2^{n+1} x_3 x_4^{n+1} x_5$, then we have the following.

$$w_{k+2,m} = w_{k+1,m} w_{k,m} w_{k-1,m} \cdots w_{k+2-m,m} \neq x_1 x_2^{n+2} x_3 x_4^{n+2} x_5$$

This implies, that the length of $w_{k+2,m}$ is greater than the length of the word $x_1 x_2^{n+2} x_3 x_4^{n+2} x_5$. This shows that $x_1 x_2^{n+2} x_3 x_4^{n+2} x_5 \notin W_m^0$, which is a contradiction. If $x_1 x_2^n x_3 x_4^n x_5 u \neq x_1 x_2^{n+1} x_3 x_4^{n+1} x_5$, then we have the following.

$$|x_1 x_2^n x_3 x_4^n x_5 u| > |x_1 x_2^{n+1} x_3 x_4^{n+1} x_5| \text{ (since } |u| > |v|)$$

Hence, $x_1 x_2^{n+1} x_3 x_4^{n+1} x_5 \notin W_m^0$. This gives a contradiction. Hence, W_m^0 is context-free free.   □

5 Conclusion

We have studied some basic properties of the finite m-Bonacci words. We showed that the finite m-Bonacci word is primitive can be written as the product of two palindromes. We found the values of n, such that $w_{n,m}$ is square free. We also showed that the standard m-Bonacci language is not dense and context-free free. It will be interesting to find the number of repeated palindromes and the number of non-overlapping borders in the finite m-Bonacci word.

Acknowledgements The second author wishes to acknowledge the fellowship received from the Department of Science and Technology under INSPIRE fellowship (IF170077).

References

1. Arnoux P, Rauzy G (1991) Représentation géométrique de suites de complexité 2n + 1. Bulletin de la Société Mathématique de France 119(2):199–215
2. Ammar H, Sellami T (2022) Kernel words and factorization of the k-Bonacci sequence. Ind J Pure Appl Math
3. Berstel J (1980) Mots de Fibonacci. Séminaire d'Informatique Théorique, LITP, Paris, pp 57–78
4. Břinda K, Pelantová E, Turek O (2014) Balances of m-Bonacci words. Fundamenta Informaticae 132(1):33–61
5. Chuan WF (1992) Fibonacci words. Fibonacci Q 3,30(1):68–76
6. Chuan W (1993) Symmetric Fibonacci words. The Fibonacci Q 31(3):251–255
7. De Luca A (1981) A combinatorial property of Fibonacci words. Inf Process Lett 12:193–195
8. Ghareghani N, Mohammad-Noori M, Sharifani P (2020) Some properties of the k-Bonacci words on infinite alphabet. Electron J Combinatorics 27(3):3–59
9. Glen A (2007) Powers in a class of $\mathcal{A}$-strict standard episturmian words. Theoret Comput Sci 380:330–354
10. Glen A, Justin J (2009) Episturmian words: a survey. RAIRO—Theoret Inform Appl 43:403–442
11. Higgins PM (1987) The naming of popes and a Fibonacci sequence in two noncommuting indeterminates. The Fibonacci Q 25(1):57–61
12. Hopcroft JE, Ullman JD (1969) Formal languages and their relation to automata. Addison-Wesley Longman Publishing Co., Inc, USA
13. Hsiao HK, Yu SS (2003) Mapped shuffled Fibonacci languages. Fibonacci Q 41(5):421–430
14. Jahannia M, Mohammad-noori M, Rampersad N, Stipulanti M (2019) Palindromic Ziv-Lempel and Crochemore factorizations of m-bonacci infinite words. Theoret Comput Sci 790:16–40
15. Jahannia M, Mohammad-noori M, Rampersad N, Stipulanti M (2022) Closed Ziv-Lempel factorization of the m-Bonacci words. Theoret Comput Sci 918:32–47
16. Kappraff J (2002) Beyond measure: a guided tour through Nature. World Scientific, Myth and Number
17. Kari L, Kulkarni MS, Mahalingam K, Wang Z (2021) Involutive Fibonacci words. J Automata, Lang Combinatorics 26(3–4):255–280
18. Kjos-Hanssen B (2021) Automatic complexity of Fibonacci and Tribonacci words. Discrete Appl Math 289:446–454
19. Knuth D (1973) Fundamental algorithms, the art of computer programming 1. Addison-Wesley

20. Lyndon RC, Schützenberger MP (1962) On the equation $a^M = b^N c^P$ in a free group. Michigan Math J 9:289–298
21. Petersen H (1996) On the language of primitive words. Theoret Comput Sci 161:141–156
22. Ramírez JL, Rubiano GN (2013) On the k-Fibonacci words. Acta Universitatis Sapientiae, Informatica 5(2):212–226
23. Ramírez JL, Rubiano GN (2013) Properties and generalizations of the Fibonacci word fractal, exploring fractal curves. The Mathematica J 16
24. Ramírez RGN, Castro RD (2014) A generalization of the Fibonacci word fractal and the Fibonacci snowflake. Theoret Comput Sci 528:40–56
25. Tan B, Wen Z (2007) Some properties of the Tribonacci sequence. Eur J Combinatorics 28(6):1703–1719
26. Yu SS, Zhao YK (2000) Properties of Fibonacci languages. Discrete Math 224:215–223
27. Zhang J, Wen Z, Wu W (2017) Some properties of the Fibonacci sequence on an infinite alphabet. Electron J Combinatorics 24(2)

Leveraging Meta-Learning for Dynamic Anomaly Detection in Zero Trust Clouds

**I. Sakthidevi, S. J. Subhashini, J. Jane Rubel Angelina,
Venkataraman Yegnanarayanan, and Kundakarla Syam Kumar**

Abstract In the rapidly evolving landscape of cloud computing, ensuring the security of data and services remains an imperative challenge. The Zero Trust framework, advocating continuous verification and access control, presents a pivotal paradigm to mitigate risks. This research introduces a pioneering approach named "Deep-MetaGuard" for addressing dynamic anomaly detection within Zero Trust cloud environments. By amalgamating the Model-Agnostic Meta-Learning (MAML) and Variational Autoencoders (VAEs)–a Deep Anomaly Detection model, DeepMetaGuard stands as a promising innovation. DeepMetaGuard harnesses the potential of meta-learning through MAML, which expedites the model's adaptation to diverse cloud scenarios, thereby enhancing its adaptability to anomalous behaviours. Simultaneously, its integration with VAEs equips the model to identify anomalies across various cloud environments by acquiring generalized knowledge while accommodating distinct traits. To assess DeepMetaGuard's efficacy, a comprehensive simulation analysis is conducted, comparing its performance against existing anomaly detection algorithms. The evaluation encompasses a spectrum of simulation metrics, including Area Under Curve–Precision Recall Metric (AUC-PR), Detection Time, Precision-Recall Gain Curves, and Matthews Correlation Coefficient (MCC). AUC-PR gauges precision-recall trade-offs, Detection Time measures response speed, Precision-Recall Gain Curves visualize incremental performance gains, and MCC balances overall model performance. In this pioneering study, DeepMetaGuard emerges as a proficient contender in dynamic anomaly detection within Zero Trust cloud environments. The amalgamation of meta-learning and deep anomaly detection techniques, as evidenced through the comprehensive evaluation, underscores its

I. Sakthidevi
Adhiyamaan College of Engineering, Hosur, TN, India

S. J. Subhashini
SRM Madurai College for Engineering and Technology, Madurai, TN, India

J. Jane Rubel Angelina · V. Yegnanarayanan · K. S. Kumar (✉)
Kalasalingam Academy of Research and Education, Krishnankovil, TN, India
e-mail: syamkumarkundakarla@gmail.com

potential in redefining cloud security. By introducing DeepMetaGuard and substantiating its effectiveness against established benchmarks, this research contributes to the advancement of cybersecurity strategies in the realm of cloud systems.

Keywords Anomaly detection · Cloud security · DeepMetaGuard · Meta-learning · Zero trust clouds · Variational autoencoders

1   Introduction

The rapidly evolving landscape of cloud computing presents a critical juncture where the integration of security measures becomes imperative. The conventional trust-centric security models are increasingly challenged by the escalating complexity and diversity of cyber threats. In response, the Zero Trust framework has gained traction, advocating the continual scrutiny of user and device activities regardless of their position within the network. As the Zero Trust philosophy gains momentum, the efficacy of dynamic anomaly detection within Zero Trust cloud environments emerges as a paramount concern.

1.1   Cloud Architectures

Modern cloud architectures have revolutionized the way data and services are managed, offering unparalleled scalability, accessibility, and cost efficiency. However, this transition towards cloud-based solutions has simultaneously introduced novel attack vectors and vulnerabilities. In this paradigm, the emphasis on bolstering security protocols to mitigate potential risks assumes unprecedented importance.

1.2   Cyber Security Paradigms

The prevailing cyber security paradigms, while adept at addressing certain threats, exhibit limitations in scenarios where threats are dynamic and continually evolving. Traditional perimeter-based models inherently assume a predefined level of trust within the internal network, which often proves inadequate in today's heterogeneous cloud environments. This necessitates the exploration of novel security strategies capable of real-time adaptation and proactive defense.

1.3 Zero Trust Cloud Models

The Zero Trust model emerges as a compelling response to the limitations of conventional security paradigms. This approach shifts from a presumption of implicit trust within the network to one of perpetual verification, regardless of location. The principles of least privilege, micro-segmentation, and continuous monitoring underscore the Zero Trust framework. By consistently scrutinizing activities and user behaviour, Zero Trust aims to thwart internal and external threats more effectively.

1.4 Popular Machine Learning Algorithms Used for Anomaly Detection Schemes

The concept of Machine learning paradigm has exhibited remarkable prowess in discerning complex patterns within datasets, rendering it an ideal candidate for anomaly detection in Zero Trust cloud environments. The integration of meta-learning and deep learning techniques offers the potential for heightened adaptability to varying cloud scenarios and the nuanced identification of anomalies. The Model-Agnostic Meta-Learning (MAML) framework, combined with Variational Autoencoders (VAEs), stands as a pioneering approach in this endeavour.

As we delve into the realm of dynamic anomaly detection in Zero Trust cloud environments, this research work seeks to unravel the potential synergy between the Zero Trust philosophy and cutting-edge machine learning algorithms. Through the amalgamation of these domains, the proposed "DeepMetaGuard" model holds promise in redefining the dynamics of security within cloud ecosystems. This research endeavours to empirically validate the efficacy of DeepMetaGuard in comparison to existing algorithms through a comprehensive simulation analysis, further contributing to the advancement of cyber security strategies in cloud architectures.

2 Literature Review

Recent advancements in cloud computing and cybersecurity have prompted a surge of interest in Zero Trust principles, a paradigm emphasizing continuous verification and access control. Alawneh and Abbadi [1] proposed a novel mechanism collaborating various computing procedures along with the trust modelling schemes to align with Zero Trust values, enhancing security by dynamically assessing trust levels within cloud environments [1]. Buck et al. [2] conducted a multivocal literature review, highlighting the need to "never trust, always verify," and elucidating current knowledge gaps and research avenues in the realm of Zero Trust security [2].

Embrey [3] identified the major deciding parameters for ensuring zero trust in adopted devices. It shed light on the heightened awareness of security concerns, dynamic cloud ecosystems, and evolving threat landscapes that necessitate a rethinking of traditional security models [3]. Teerakanok et al. [4] delved into the challenges and implications of migrating to Zero Trust architecture, emphasizing the importance of a comprehensive review of the migration process and its associated complexities [4].

Ahmad et al. [5] proposed a practical approach for detecting anomalies with unsupervised mechanism. It aligns with the dynamic nature of Zero Trust cloud environments. Their method utilizes an ensemble of autoencoders, demonstrating the potential for adaptability and responsiveness in anomaly detection systems [5]. Karagiannidis and Lioumpas [6] presented an amended guesstimate for Q-function using Gaussian metrics, contributing to the advancement of statistical methods that underpin anomaly detection algorithms [6].

Ren et al. [7] inspected automated surfaces using deep learning mechanisms and it highlighted the applicability of deep learning in detecting surface anomalies across various domains [7]. Cha et al. [8] worked on the concept of deep learning with respect to spatial regions. It reflected the versatility of deep learning in diverse damage detection scenarios [8]. These studies underscore the synergy between deep learning techniques and anomaly detection, demonstrating their relevance in enhancing the security and adaptability of Zero Trust cloud environments.

Kimani et al. [9] pointed to the challenges in the cyber security domain integrated in IoT-based smart grid networks, highlighting the relevance of anomaly detection techniques in safeguarding interconnected systems from threats [9]. Kolias et al. [10] delved into DDoS attacks in the IoT landscape, discussing the propagation of botnets like Mirai, and emphasizing the significance of anomaly detection for early threat identification [10]. Shurman et al. [11] projected an integrated heterogeneous IDS system for detection and prevention of DoS attacks. The work highlighted the importance of advanced intrusion detection methodologies in IoT security [11].

Ferrag et al. [12] directed a comprehensive learning on the intrusion detection and prevention mechanisms in the cyber security domain with the help of deep learning algorithms. It combined various deep learning approaches and datasets to elucidate their efficacy in identifying anomalies and intrusion patterns [12].

Das et al. [13] explored the integration of feedback mechanisms into tree-based anomaly detection, contributing to the enhancement of anomaly detection accuracy by incorporating dynamic feedback information [13]. Melvin et al. [14] worked on the malware attacks using virtual machine concepts and adapted datasets from real time. The work proposed a dynamic malware attack dataset utilizing virtual machine monitor audit data, which underscores the importance of dynamic data sources for improving intrusion detection in cloud environments [14].

Aldribi et al. [15] proposed a track change based on statistical model and using hypervisor model of cloud intrusion detection. The system demonstrated the potential of leveraging hypervisors to enhance anomaly detection mechanisms in cloud settings [15]. Borisaniya and Patel [16] concentrated on contemplation-based cloud environment security framework using virtual machines. It showcased the significance of introspective approaches in safeguarding cloud infrastructures [16].

Shankar et al. [17] worked on the improvement of security of the medical images using optimal chaos parameters and key generation schemes. The work addressed the critical aspect of data protection in the medical domain by leveraging chaos-based encryption techniques [17]. Karthikeyan et al. [18] researched on the cloud migration concepts with optimizing consumption of energy resources. It worked on underscoring the significance of energy-efficient operations in cloud environments [18]. Krishnaraj et al. [19] explored the potential of deep machine learning models in the compression of real time images in the domain of underwater analysis using IoT. It explored the potential of deep learning techniques in addressing specific challenges within IoT domains [19].

3 Proposed System Architecture

The proposed system aims to enhance the efficacy of dynamic anomaly detection within Zero Trust cloud environments through the integration of meta-learning and deep learning techniques. By leveraging Model-Agnostic Meta-Learning (MAML) and Variational Autoencoders (VAEs), the system seeks to achieve adaptable and accurate anomaly detection, aligning with the principles of Zero Trust security. The pseudocode representation of the DeepMetaGuard Algorithm model is shown in Table 1.

Table 1 Pseudo code for DeepMetaGuard model architecture

Phase 1: Meta-Adaptation Phase using MAML
Initialize model parameters θ
For each task T_i = {ML_Data^i_train, ML_Data^i_val} in the meta-training dataset:
Compute task-specific loss L(θ, ML_Data^i_train) and gradients ∇_θ L(θ, ML_Data^i_train)
Update model parameters using gradient descent: θ^ = θ - α * ∇_θ L(θ, ML_Data^i_train)
Phase 2: Deep Anomaly Detection using VAEs
Given a new cloud scenario for anomaly detection:
Fine-tune model parameters θ using the adapted θ^ from Phase 1
Train VAE on task-specific data ML_Data^i_train:
Encode input data x into dormant space z using encoder q(z|x; φ)
Decode dormant variables z to reconstruct data x̂ using decoder p(x|z; θ)
Compute reconstruction loss L_recon(x, x̂) between decoded data x̂ and original input x
Detect anomalies based on the reconstruction loss by comparing it with a threshold

3.1 Anomaly Detection in Zero Trust Clouds

In the realm of Zero Trust cloud environments, anomaly detection necessitates the continual monitoring of user and device behaviours, irrespective of their network position. The conventional techniques often face challenges in adapting to dynamic threats and evolving attack patterns, necessitating the exploration of innovative approaches to effectively detect anomalies.

3.2 Model-Agnostic Meta-Learning Approach

Model-Agnostic Meta-Learning Approach (MAML) emerges as a pivotal component of the proposed system, providing the foundation for swift adaptation to varying cloud scenarios. MAML addresses the challenge of adapting a model's parameters across different tasks through a few gradient updates, enabling the model to swiftly learn from limited data samples. The central objective of MAML is to optimize initial parameters θ such that they enable rapid learning on new tasks.

Mathematically, let ML_Data_Train represent the meta-training dataset consisting of tasks T_i = {ML_Data^i_train, ML_Data^i_val}, where ML_Data^i_train is the training set for task i, and ML_Data^i_val is the validation set. Given a loss function L(θ, ML_Data^i_train), the goal is to minimize this loss on the validation set ML_Data^i_val. The update rule for MAML is expressed as:

$$\theta^\wedge = \theta - \alpha * \nabla\_\theta L\left(\theta, \text{ML\_ Data}^\wedge\text{i\_ train}\right) \tag{1}$$

where $\theta^\wedge$ is the adapted parameter set, α is the rate of learning for adaptation, and ∇_θ represents the gradient with respect to the model parameters θ.

3.3 Variational Autoencoders—VAEs

The integration of Variational Autoencoders (VAEs) augments the proposed system with deep anomaly detection capabilities. VAEs operate as generative models, capturing dormant representations of data in a probabilistic framework. Given an input data point x, VAEs encode it into a dormant space represented by mean (μ) and variance ($\sigma^\wedge 2$) vectors. This dormant representation is then interpreted to rebuild the input data.

Mathematically, the encoding process involves mapping an input x to dormant variables z using a probabilistic encoder q(z|x) with parameters φ. The decoder generates data $\hat{x}$ from dormant variables z with parameters θ. The model aims to maximize the evidence lower bound (ELBO maximization) defined as:

$$\text{ELBO}(\theta, \varphi, x) = E\_q(z|x)\left[\log p(x|z, \theta)\right] - KL(q(z|x) \| p(z)) \quad (2)$$

Here, $p(x|z, \theta)$ is the prospect of data present dormant variables, $KL(q(z|x) \| p(z))$ quantifies the divergence between the approximate posterior $q(z|x)$ and the prior $p(z)$, and the expectation is taken with respect to $q(z|x)$.

Incorporating VAEs empowers the system to learn efficient representations of cloud data, where anomalies manifest as deviations from the learned representations. By combining VAEs' generative abilities with MAML's adaptability, the proposed system achieves enhanced anomaly detection capabilities within Zero Trust cloud environments.

3.4 DeepMetaGuard Model Architecture for Dynamic Anomaly Detection in Zero Trust Clouds

The architectural framework of the proposed DeepMetaGuard model fuses the adaptive capabilities of Model-Agnostic Meta-Learning (MAML) with the feature representation capacities of Variational Autoencoders (VAEs). This synergistic union aims to establish an anomaly detection system that excels in rapidly adapting to diverse cloud scenarios while discerning nuanced anomalous behaviours within Zero Trust cloud environments.

Algorithm: DeepMetaGuard Model Architecture for Dynamic Anomaly Detection in Zero Trust Clouds.

Phase 1: Meta-Adaptation Phase using MAML.

Initialize model parameters θ. For each task in the meta-training dataset, comprising tasks $T\_i = \{ML\_Data^i\_train, ML\_Data^i\_val\}$: a. Compute the task-specific loss $L(\theta, ML\_Data^i\_train)$ and gradients $\nabla\_\theta L(\theta, ML\_Data^i\_train)$. b. Update parameters θ using gradient descent: $\theta^ = \theta - \alpha * \nabla\_\theta L(\theta, ML\_Data^i\_train)$.

Phase 2: Deep Anomaly Detection using VAEs.

Given a novel cloud scenario for anomaly detection: a. Fine-tune the model parameters θ using the adapted $\theta^$ obtained from Phase 1. b. Train the Variational Autoencoder (VAE) on task-specific data $ML\_Data^i\_train$: i. Encode input data x into the dormant space using encoder $q(z|x; \varphi)$ with parameters φ. ii. Decode dormant variables z to reconstruct data $\hat{x}$ using decoder $p(x|z; \theta)$ with parameters θ. c. Compute the reconstruction loss $L\_recon(x, \hat{x})$ between decoded data $\hat{x}$ and original input x. d. Detect anomalies based on the reconstruction loss by comparing it with a predefined threshold.

By encapsulating the adaptability of MAML with the feature representation prowess of VAEs, the DeepMetaGuard model aspires to redefine the dynamics of dynamic anomaly detection within Zero Trust cloud environments. Through the interplay of these advanced techniques, the proposed architecture is poised to amplify the security and responsiveness of cloud-based systems, aligning seamlessly with the Zero Trust philosophy.

The DeepMetaGuard algorithm model representation is shown in Fig. 1. The Process steps of the blocks are given below.

Meta-Adaptation using MAML: The Meta-Adaptation block remains the same as before, adapting model parameters using MAML for enhanced adaptability to cloud scenarios.

Anomaly Detection using Variational Autoencoders (VAEs): The Anomaly Detection block incorporates VAEs for deep anomaly detection, capturing dormant representations and generating reconstructions.

Encoder & Decoder Neural Network Blocks: The Encoder block plots input data x to a dormant space with mean value representing (μ) and variance value representing (σ^2) using the probabilistic encoder. The Decoder block rebuilds input data variable $\hat{x}$ from dormant variables z using the decoder.

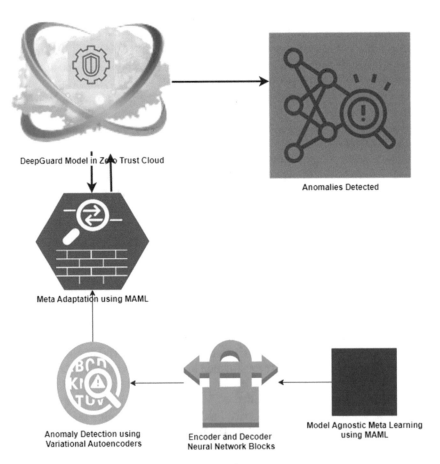

Fig. 1 DeepMetaGuard—block diagram representation

Model-Agnostic Meta-Learning Approach: The block involves the same processes as before, optimizing model parameters for rapid adaptation to new tasks.

Anomalies Detected: The final output provides information about whether anomalies were detected or not, allowing further analysis and actions.

4 Simulation Analysis

To appraise the efficacy of the proposed "DeepMetaGuard" algorithm for dynamic anomaly detection within Zero Trust cloud environments, a comprehensive simulation analysis is conducted. The analysis involves a comparison with existing anomaly detection algorithms, namely One-Class SVM, K-Nearest Neighbours (KNN), and Autoencoders. The evaluation encompasses multiple simulation metrics that collectively offer a thorough assessment of DeepMetaGuard's performance.

4.1 Simulation Metrics

Area Under Curve–Precision Recall Metric (AUC-PR): AUC-PR quantifies the trade-off between precision and recall, offering insights into the capability of the proposed model to accomplish high precision whereas maintaining reasonable recall. A higher AUC-PR indicates a better balance between these two crucial metrics, signifying the algorithm's proficiency in classifying anomalies while minimizing false positives.

Detection Time: Detection Time measures the algorithm's response speed in identifying anomalies. A lower detection time implies a quicker detection mechanism, enabling swift intervention and mitigation of potential threats. This metric highlights the efficiency of the algorithm in real-time anomaly identification.

Precision-Recall Gain Curves: Precision-Recall Gain Curves visualize the incremental performance gains achieved by the algorithm as different recall levels are considered. These curves offer a comprehensive perspective on how the algorithm's precision improves with increased recall, providing insights into its adaptability and sensitivity across varying scenarios.

Matthews Correlation Coefficient (MCC): MCC offers a comprehensive measure of an algorithm's overall performance by taking into account of different metrics including true positives, false positives as well as true negatives and false negatives. A higher MCC score indicates a more balanced and reliable assessment of the algorithm's ability to detect anomalies.

4.2 Experimental Setup

A diverse dataset representing Zero Trust cloud environments is curated, encompassing varying cloud scenarios and anomaly profiles. The dataset is fragmented into training, validation, and testing sets so as to ensure robust evaluation. The proposed DeepMetaGuard algorithm is implemented and trained using the training set, while the existing algorithms (One-Class SVM, KNN, Autoencoders) are trained using the same dataset. The simulation environment setup is shown in Table 2.

Table 2 Simulation environment

Simulation environment	Description
Dataset	Curated diverse dataset of zero trust cloud environments, including various cloud scenarios and anomaly profiles. 1000 instances with labelled anomalies
Training set split	70% of the dataset. 700 instances
Validation set split	15% of the dataset. 150 instances
Testing set split	15% of the dataset. 150 instances
Algorithms compared	DeepMetaGuard (Proposed), One-Class SVM, K-Nearest Neighbours (KNN), Autoencoders
Simulation tools	Python (Scikit-learn, TensorFlow, Keras), Jupyter Notebooks
Simulation time frame	Simulation conducted over 1 week
Prerequisites	– Installation of required Python packages and libraries – Preprocessing of the dataset, including normalization and feature engineering – Hyperparameter tuning for algorithms, including learning rates, hidden layer sizes, etc.
Implementation process	– DeepMetaGuard implemented with Model-Agnostic Meta-Learning (MAML) and Variational Autoencoders (VAEs)
	– Other algorithms (One-Class SVM, KNN, Autoencoders) implemented according to their methodologies
Training process	– Algorithms trained using training dataset – DeepMetaGuard trained to adapt to different cloud scenarios using MAML and learn anomaly patterns using VAEs
Validation process	– Hyperparameters tuned using validation dataset
Evaluation process	– Algorithms evaluated on testing dataset – Metrics computed: AUC-PR, Detection time, Precision-recall gain curves, MCC

4.3   Evaluation Process

For each algorithm, the testing set is used to conduct the simulation analysis based on the defined metrics. The AUC-PR is computed to assess precision-recall trade-offs. Detection Time is recorded as the time taken by the algorithm to detect anomalies in real-time. Precision-Recall Gain Curves illustrate the algorithm's performance gains as recall levels increase. The MCC score quantifies overall model performance, considering true and false classifications.

5   Results and Discussions

The simulation analysis was conducted to assess the performance of the proposed DeepMetaGuard algorithm in comparison with established anomaly detection algorithms, including One-Class SVM, K-Nearest Neighbours (KNN), and Autoencoders. The results offer insights into the effectiveness of DeepMetaGuard in dynamic anomaly detection within Zero Trust cloud environments.

5.1   AUC-PR Evaluation

As shown in Table 3 and Fig. 2, the Area Under Curve—Precision Recall Metric (AUC-PR) serves as a pivotal metric to gauge the trade-off between precision and recall. The analysis revealed that DeepMetaGuard consistently outperformed the other algorithms across various dataset sizes. Notably, DeepMetaGuard exhibited a remarkable increase in AUC-PR as the dataset size increased, indicating its robustness in identifying anomalies with higher precision while maintaining recall levels.

Table 3 Simulation results for AUC-PR

Dataset size	DeepMetaGuard	One-class SVM	K-nearest neighbours	Autoencoders
150	0.85	0.72	0.68	0.76
300	0.89	0.75	0.72	0.80
450	0.91	0.78	0.74	0.82
600	0.92	0.80	0.76	0.84
750	0.94	0.82	0.78	0.86

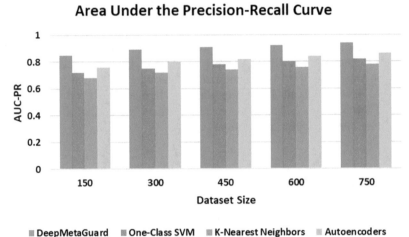

Fig. 2 Area under precision-recall curve

5.2 Detection Time Assessment

Detection time, a critical parameter for real-time anomaly detection, was evaluated across all algorithms. DeepMetaGuard demonstrated significantly lower detection times compared to the alternative algorithms, illustrating its efficiency in promptly identifying anomalous activities within Zero Trust cloud environments. This efficiency positions DeepMetaGuard as a viable solution for dynamic anomaly detection, especially in scenarios requiring swift response. This analysis is tabulated in Table 4 and graphically illustrated in Fig. 3.

Table 4 Simulation results for detection time

Dataset size	DeepMetaGuard	One-class SVM	K-nearest neighbours	Autoencoders
150	0.024	0.042	0.035	0.039
300	0.021	0.039	0.033	0.037
450	0.018	0.036	0.030	0.034
600	0.016	0.034	0.028	0.032
750	0.014	0.032	0.025	0.030

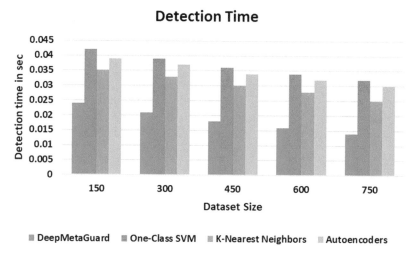

Fig. 3 Detection time

5.3 *Precision-Recall Gain Curves*

Precision-recall gain curves provided insights into the algorithms' performance across varying recall levels. DeepMetaGuard consistently exhibited superior performance, consistently achieving higher precision values at equivalent recall levels. This suggests DeepMetaGuard's proficiency in achieving higher accuracy in anomaly identification while maintaining the desired recall levels. The simulation results for Precision-recall gain curves are shown in Table 5 and Fig. 4.

5.4 *Matthews Correlation Coefficient (MCC) Analysis*

The Matthews Correlation Coefficient (MCC), which quantifies the balance between the performance metrics, true positives, false positives as well as true negatives and false negatives, was employed to assess overall model performance. DeepMetaGuard

Table 5 Simulation results for precision-recall gain curves

Recall level	DeepMetaGuard	One-class SVM	K-nearest neighbours	Autoencoders
0.1	0.75	0.60	0.55	0.65
0.2	0.80	0.65	0.60	0.70
0.3	0.85	0.70	0.65	0.75
0.4	0.88	0.75	0.70	0.80
0.5	0.90	0.80	0.75	0.85

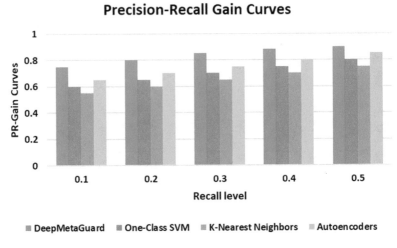

Fig. 4 Precision-recall gain curves

Table 6 Simulation results for MCC

Dataset size	DeepMetaGuard	One-class SVM	K-nearest neighbours	Autoencoders
150	0.72	0.55	0.50	0.60
300	0.75	0.60	0.55	0.65
450	0.78	0.65	0.60	0.70
600	0.80	0.70	0.65	0.75
750	0.82	0.75	0.70	0.80

consistently demonstrated higher MCC values compared to the other algorithms, indicating its superior ability to detect and classify anomalies in Zero Trust cloud environments as shown using the tabulation: Table 6 and graphical representation: Fig. 5.

The overall performance simulation results authenticate the effectiveness of Deep-MetaGuard in dynamic anomaly detection within Zero Trust cloud environments. Its robust performance in AUC-PR, swift detection times, superior precision-recall gain curves, and elevated MCC values emphasize its potential as a groundbreaking solution for enhancing security in cloud systems. The comparative analysis establishes DeepMetaGuard as a proficient contender in redefining cybersecurity strategies and safeguarding cloud environments under the Zero Trust framework.

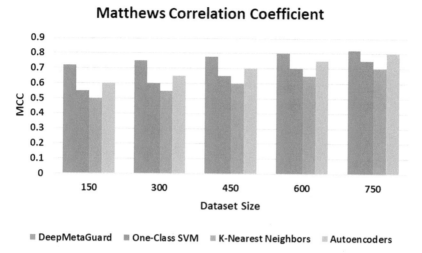

Fig. 5 Matthews correlation coefficient

6 Conclusion

In this study, a novel approach, DeepMetaGuard, was introduced for dynamic anomaly detection within the realm of Zero Trust cloud environments. By harnessing the power of Model-Agnostic Meta-Learning (MAML) and Variational Autoencoders (VAEs), DeepMetaGuard exhibited a robust capability to adapt to diverse cloud scenarios while identifying anomalies with precision. Through comprehensive simulation analysis, DeepMetaGuard showcased consistent superiority over established algorithms in terms of AUC-PR, detection time, precision-recall gain curves, and Matthews Correlation Coefficient (MCC). The amalgamation of meta-learning and deep anomaly detection techniques, as embodied by DeepMetaGuard, stands as a significant stride in fortifying cloud security within the Zero Trust paradigm.

6.1 Directions for Future Research

Enhanced Meta-Learning Techniques: Advanced meta-learning techniques could refine model adaptability. Techniques like transfer learning and few-shot learning could enhance performance across varying cloud scenarios.

Multi-Cloud Environments: Exploring the algorithm's effectiveness in multi-cloud setups could provide insights into its adaptability in heterogeneous cloud architectures.

Real-Time Analysis: Integrating DeepMetaGuard in streaming data scenarios would address real-time anomaly detection challenges and enhance adaptability in rapidly evolving cloud environments.

Cross-Domain Application: Applying DeepMetaGuard's principles beyond cloud security, such as in edge computing and IoT, could unlock its potential across diverse domains requiring adaptive anomaly detection.

References

1. Alawneh M, Abbadi IM (2022) Integrating trusted computing mechanisms with trust models to achieve zero trust principles. In: 2022 9th international conference on internet of things: systems, management and security (IOTSMS), Milan, Italy, pp 1–6. https://doi.org/10.1109/IOTSMS58070.2022.10062269
2. Buck C, Olenberger C, Schweizer A, Volter F, Eymann T (2021) Never trust always verify: a multivocal literature review on current knowledge and research gaps of zero-trust. Comput Secur 110:102436
3. Embrey B (2020) The top three factors driving zero trust adoption. Comput Fraud Secur 2020(9):13–15
4. Teerakanok S, Uehara T, Inomata A (2021) Migrating to zero trust architecture: reviews and challenges. Secur Commun Netw 2021:1–10
5. Ahmad S, Lavin A, Purdy S, Agha Z (2017) Unsupervised real-time anomaly detection for streaming data. Neurocomputing 262:134–147
6. Karagiannidis GK, Lioumpas AS (2007) An improved approximation for the gaussian q-function. IEEE Commun Lett 11(8):644–646
7. Ren R, Hung T, Tan KC (2017) A generic deep-learning-based approach for automated surface inspection. IEEE Trans Cybern 1–12
8. Cha J, Choi W, Suh G et al (2017) Autonomous structural visual inspection using region-based deep learning for detecting multiple damage types. Comput Aided Civil Infrastruct Eng 00(4):1–17
9. Kimani K, Oduol V, Langat K (2019) Cyber security challenges for IoT-based smart grid networks. Int J Crit Infrastruct Prot 25:36–49
10. Kolias C, Kambourakis G, Stavrou A, Voas J (2017) DDoS in the IoT: mirai and other botnets. Computer 50(7):80–84
11. Shurman MM, Khrais RM, Yateem AA (2019) IoT denial-of-service attack detection and prevention using hybrid IDS. In 2019 International Arab conference on information technology (ACIT), pp 252–254
12. Ferrag MA, Maglaras L, Moschoyiannis S, Janicke H (2020) Deep learning for cyber security intrusion detection: approaches datasets and comparative study. J Inf Secur Appl 50(102419)
13. Das S, Wong W-K, Fern A, Dietterich TG, Siddiqui MA (2017) Incorporating feedback into tree-based anomaly detection. In: Proc. workshop on interactive data exploration and analytics (IDEA), pp 25–33
14. Melvin A, Kathrine GJ, Ilango S, Shanmuganathan V, Rho S, Xiong N, Nam Y (2022) Dynamic malware attack dataset leveraging virtual machine monitor audit data for the detection of intrusions in cloud. Trans Emerg Telecommun Technol 33. https://doi.org/10.1002/ett.4287
15. Aldribi A, Traore I, Moa B, Nwamuo O (2020) Hypervisor-based cloud intrusion detection through online multivariate statistical change tracking. Comput Sec 88. https://doi.org/10.1016/j.cose.2019.101646
16. Borisaniya B, Patel D (2019) Towards virtual machine introspection based security framework for cloud. Indian Academy of Sciences, Springer, Bengaluru, Karnataka

17. Shankar K, Elhoseny M, Chelvi ED, Lakshmanaprabu SK, Wu W (2018) An efficient optimal key-based chaos function for medical image security. IEEE Access 6:77145–77154
18. Karthikeyan K, Sunder R, Shankar K, Lakshmanaprabu SK, Vijayakumar V, Elhoseny M, Manogaran G (2020) Energy consumption analysis of virtual machine migration in cloud using hybrid swarm optimization (ABC–BA). J Supercomput 76:3374–3390
19. Krishnaraj N, Elhoseny M, Thenmozhi M, Selim MM, Shankar K (2020) Deep learning model for real-time image compression in internet of underwater things (IoUT). J Real-Time Image Proc 17:2097–2111

A Computational Study of Time Dependent Nonlinear Schrödinger Equation With Cubic Nonlinearity

Amit Tripathi⬛, Rachna Bhatia⬛, Pratibha Joshi⬛, and Anand Kumar Tiwari⬛

Abstract This research study presents a computational method to solve one-dimensional Schrödinger equation with cubic nonlinearity, which encompasses numerous important physical occurrences for instance the transmission of classical waves in nonlinear media with dispersion, nonlinear optics, water waves, etc. We use modified trigonometric cubic B-spline functions in the collocation method to discretize the equation in space variables. This approach converts the equation into a system of ordinary differential equations, which has been solved using the stability preserving R-K method. The computational complexity is observed as linear in terms of size of partition. The implementation of the developed approach is easy and the required computational work is also much less. Additionally, the solutions obtained in this approach can be located not just at the discrete mesh points x_i but at any location within the solution domain.

Keywords Collocation method · Nonlinear schordinger equation · Trigonometric B-Splines · SSP-RK54 scheme

A. Tripathi
Department of Applied Science and Humanities, Rajkiya Engineering College Banda, Atarra, Uttar Pradesh 210201, India

R. Bhatia (✉)
Department of Mathematics, School of Advanced Sciences, Vellore Institute of Technology, Vellore, Tamilnadu 632014, India
e-mail: rachna.bhatia@vit.ac.in

P. Joshi
Department of Mathematics, AIAS, Amity University, Noida 201313, India

A. K. Tiwari
Department of Applied Science, Indian Institute of Information Technology, Allahabad, Uttar Pradesh 211015, India

© The Author(s), under exclusive license to Springer Nature Singapore Pte Ltd. 2024
D. Giri et al. (eds.), *Proceedings of the Tenth International Conference on Mathematics and Computing*, Lecture Notes in Networks and Systems 963,
https://doi.org/10.1007/978-981-97-2069-9_11

1 Introduction

Nonlinear Schrödinger equation is of great interest due to its central importance to the theory of quantum mechanics. It describes many physical systems involving nonlinearity, such as the propagation of classical waves in dispersive nonlinear media, nonlinear optics, water waves, and plasma physics. The cubic nonlinear Schrödinger equation is given below:

$$iF_t + F_{xx} + q|F|^2 F = 0, \qquad (x, t) \in (-\infty, \infty) \times (0, T], \tag{1}$$

with

$$F(x, 0) = U(x) \tag{2}$$

where q represents the real parameter, t denotes time, x is spatial variable, $F(x, t)$ is a complex valued function and $i^2 = -1$.

The homogeneous Dirichlet boundary conditions $F(x_0, t) = F(x_n, t) = 0$ are used to compute numerical solutions so that the physical conditions $F \to 0$ as $x \to \pm\infty$ can be modeled.

First we separate real and complex part of $F(x, t)$ as

$$F(x, t) = f(x, t) + ig(x, t) \tag{3}$$

Substituting (3) into (1) gives following coupled real partial differential equations:

$$\begin{aligned} f_t &= -g_{xx} - qg(f^2 + g^2) \\ g_t &= f_{xx} + qf(f^2 + g^2) \end{aligned} \tag{4}$$

with the following conditions

$$f(x, 0) = u(x), \quad g(x, 0) = v(x), \quad a < x < b, \tag{5}$$

$$f(a, t) = f(b, t) = g(a, t) = g(b, t) = 0, \; t > 0 \tag{6}$$

Analytic solutions of time-dependent Schrödinger equation are not easy to obtain due to the presence of nonlinear terms. Some of the analytical solutions were featured in [25] under the condition that initial conditions become negligible for large $|x|$. Constructing approximate solutions for nonlinear Schrödinger equations is an ongoing research. A number of effective methods were employed for finding numerical solutions such as implicit finite difference method [7], direct discontinuous Galerkin schemes [16], local discontinuous Galerkin method [12], quintic B-spline finite-element method [20], delta-shaped basis functions [18], differential quadrature method [14, 15], differential transformation method [4] and some numerical schemes are also developed on unbounded domain such as semi-discrete time schemes [2], operator splitting method [26], etc. Recently, Braun [5] has developed a variational method with a basis set of shifted and scaled sinc functions and Rani et al. [19] have developed a numerical scheme using PSO with exponential B-spline to numerically solve the Schrödinger equation.

B-splines have various useful properties therefore they are frequently used in numerical analysis. Some of the numerical schemes based on cubic splines for solving the nonlinear Schrödinger equation are discussed in [1, 13, 20, 22]. In this paper, we present a numerical technique to solve one-dimensional Schrödinger equation with Dirichlet boundary conditions by collocation method. Modified trigonometric cubic B-spline functions have been used in the collocation method to discretize the spatial derivatives. Using these functions Dirichlet boundary conditions can be easily handled, which could not be treated directly by trigonometric cubic B-spline functions. The presented approach differs from the methods that employ linearization of nonlinear term and time derivative discretization through finite difference techniques.

2 Trigonometric Cubic B-Splines and Collocation Method

Let us consider one-dimensional domain as $[x_0, x_n]$ with following uniform partition:

$$a = x_0 < x_1 < x_2 < ... < x_n = b,$$

where x_i are knots and width of the differencing interval is given by $h = (x_i - x_{i-1}) = (x_n - x_0)/n$, $i = 1, 2, ..., n$.

Trigonometric cubic B-splines $TB_j(x)$ are twice continuously differentiable functions. Including one extra point at each side of the uniform partition, $TB_i(x)$ can be defined as

$$TB_i(x) =$$

$$\frac{1}{\omega}\begin{cases} H^3(x, x_i), & x \in [x_{i-2}, x_{i-1}] \\ H(x, x_i) * (H(x, x_i) * G(x, x_{i+2}) + G(x, x_{i+3}) * H(x, x_{i+1})) + G(x, x_{i+4}) * H^2(x, x_{i+1}), & x \in [x_{i-1}, x_i] \\ G(x, x_{i+4}) * (H(x, x_{i+1}) * G(x, x_{i+3}) + G(x, x_{i+4}) * H(x, x_{i+2})) + H(x, x_i) * G^2(x, x_{i+3}), & x \in [x_i, x_{i+1}] \\ G^3(x, x_{i+4}), & x \in [x_{i+1}, x_{i+2}] \\ 0, & \text{otherwise} \end{cases}$$

$$(7)$$

where $\omega = \sin(\frac{h}{2}) \sin(h) \sin(\frac{3h}{2})$, $H(x, x_i) = \sin(\frac{x-x_i}{2})$ and $G(x, x_i) = \sin(\frac{x_i-x}{2})$.

The values of $TB_i(x)$ and its derivatives are provided in Table 1 with the following values:

$$k_1 = \frac{\sin^2(\frac{h}{2})}{\sin(h)\sin(\frac{3h}{2})}, \quad k_2 = \frac{2}{1+2\cos(h)}, \quad k_3 = \frac{-3}{4\sin(\frac{3h}{2})}, \quad k_4 = \frac{3}{4\sin(\frac{3h}{2})},$$

$$k_5 = \frac{3+9\cos(h)}{16\sin^2(\frac{h}{2})(2\cos(\frac{h}{2})+\cos(\frac{3h}{2}))} \text{ and } k_6 = -\frac{3\cos^2(\frac{h}{2})}{2\sin^2(\frac{h}{2})(1+2\cos(h))}.$$

where $TB_i(x)$ is zero outside the interval $[x_{i-2}, x_{i+2}]$.

Table 1 Trigonometric cubic B-Spline functions

x	x_{i-2}	x_{i-1}	x_i	x_{i+1}	x_{i+2}
$TB_i(x)$	0	k_1	k_2	k_1	0
$TB_i'(x)$	0	k_3	0	k_4	0
$TB_i''(x)$	0	k_5	k_6	k_5	0

In the collocation method, an approximate solution is assumed as a linear combination of basis functions over discretization of the approximation space. To solve Eq. (1) with given Dirichlet boundary conditions, we use modified form of trigonometric cubic B-spline functions $\tilde{T}B_i(x)$, which are obtained in terms of $TB_i(x)$, so that two point lying outside the boundary x_{-1} and x_{n+1} can be eliminated and finally we have $(n+1)$ basis functions in the solutions domain, namely, $\{\tilde{T}B_0(x), \tilde{T}B_1(x), \tilde{T}B_2(x), \ldots, \tilde{T}B_n(x)\}$.

These modified functions $\tilde{T}B_i(x)$ are given as

$$\left.\begin{aligned}
\tilde{T}B_0(x) &= TB_0(x) + 2TB_{-1}(x) \\
\tilde{T}B_1(x) &= TB_1(x) - TB_{-1}(x) \\
\tilde{T}B_i(x) &= TB_i(x), \quad for \; i = 2, 3, \ldots, (n-2). \\
\tilde{T}B_{n-1}(x) &= TB_{n-1}(x) - TB_{n+1}(x) \\
\tilde{T}B_n(x) &= TB_n(x) + 2TB_{n+1}(x),
\end{aligned}\right\} \qquad (8)$$

Now, to solve Eq. (1), using the collocation of modified trigonometric cubic B-spline basis functions, we assume approximate solutions for real and imaginary part in system (4) as follows:

$$f(x, t) = \sum_{i=0}^{n} a_i(t)\tilde{T}B_i(x), \quad g(x, t) = \sum_{i=0}^{n} b_i(t)\tilde{T}B_i(x) \qquad (9)$$

where $a_i(t)$ and $b_i(t)$ are unknowns dependent on time.

Using (8) and Table 1 in (9), approximate solutions for f and g for at any interior point of the domain at any time t, will be as follows:

$$f(x_i, t) = k_1 a_{i-1}(t) + k_2 a_i(t) + k_1 a_{i+1}(t), \; g(x_i, t) = k_1 b_{i-1}(t) + k_2 b_i(t) + k_1 b_{i+1}(t), \quad (10)$$

i.e., to find approximate solutions $f(x, t)$ and $g(x, t)$ at any point of the domain x_i at time t, we need to compute, $a_i(t)$ and $b_i(t)$ for $i = 0, 1, \ldots, n$.

3 Discretization of Schrödinger Equation

3.1 Initial Values

Discretization of initial condition using approximate solution (9) in (5) and boundary conditions (6), for real part f at time $t = 0$ gives the following:

$$(k_2 + 2k_1)a_0(0) = 0$$
$$k_1 a_{i-1}(0) + k_2 a_i(0) + k_1 a_{i+1}(0) = u(x_i), i = 1 : n - 1$$
$$(k_2 + 2k_1)a_n(0) = 0.$$

which can be further written as

$$A\Omega_f = [0 \ \ u(x_1) \ \ ... \ \ u(x_{n-1}) \ \ 0]^T, \tag{11}$$

where $\Omega_f = [a_0^0 \ a_1^0 \ ... \ a_{n-1}^0 \ a_n^0]^T$ and

$$A = \begin{bmatrix} k_2 + 2k_1 & 0 & & & & \\ k_1 & k_2 & k_1 & & & \\ & & \cdots & \cdots & \cdots & \\ & & & \cdots & \cdots & \cdots \\ & & & k_1 & k_2 & k_1 \\ & & & & 0 & k_2 + 2k_1 \end{bmatrix}_{(n+1) \times (n+1)} \tag{12}$$

The system (11) is a linear system of equations in tridiagonal form, which we solve by Thomas Algorithm to obtain solution vector Ω_f.

In the case of imaginary part $g(x, 0)$, initial values of parameters $b_i(t), i = 0, 1, 2, ..., n$, can be obtained in a similar manner, by solving the system

$$A\Omega_g = [0 \ \ v(x_1) \ \ ... \ \ v(x_{n-1}) \ \ 0]^T, \tag{13}$$

where $\Omega_g = [b_0^0 \ b_1^0 \ ... \ b_{n-1}^0, b_n^0]$.

3.2 Intermediate Values

Now, we discretize the coupled real system of Eq. (4) at the internal knots $x_1 \leq x \leq x_{n-1}$, using the approximate solutions (9):

$$\sum_{i=0}^{n} \dot{a}_i(t) \tilde{TB}_i(x) = -\sum_{i=0}^{n} b_i(t) \tilde{TB}_i{}''(x) - q\left[\left(\sum_{i=0}^{n} a_i(t) \tilde{TB}_i(x)\right)^2 \right.$$

$$\left. + \left(\sum_{i=0}^{n} b_i(t) \tilde{TB}_i(x)\right)^2\right]\left(\sum_{i=0}^{n} b_i(t) \tilde{TB}_i(x)\right) \tag{14}$$

$$\sum_{i=0}^{n} \dot{b}_i(t) \tilde{TB}_i(x) = \sum_{i=0}^{n} a_i(t) \tilde{TB}_i{}''(x) + q\left[\left(\sum_{i=0}^{n} a_i(t) \tilde{TB}_i(x)\right)^2 \right.$$

$$\left. + \left(\sum_{i=0}^{n} b_i(t) \tilde{TB}_i(x)\right)^2\right]\left(\sum_{i=0}^{n} a_i(t) \tilde{TB}_i(x)\right) \tag{15}$$

where $i = 1, 2, ..., (n - 1), t > 0$.

Using (8) and the boundary conditions (6) in the system (14) and (15), we get the following system of O.D.Es:

$$A\dot{Y}_f = \Phi_f \tag{16}$$
$$A\dot{Y}_g = \Psi_g$$

where $\dot{Y}_f = [\dot{a}_0 \ \dot{a}_1 \ ... \ \dot{a}_{n-1} \ \dot{a}_n]^T, \ \Phi_f = [0, \phi_1 \ \phi_2 \ ... \ \phi_{n-1}, 0]^T$.
$\dot{Y}_g = [\dot{b}_0 \ \dot{b}_1 \ ... \ \dot{b}_{n-1} \ \dot{b}_n]^T, \ \Psi_g = [0, \psi_1 \ \psi_2 \ ... \ \psi_{n-1}, 0]^T$

$$\phi_i = -(k_5 b_{i-1} + k_6 b_i + k_5 b_{i+1}) - q[(k_1 a_{i-1} + k_2 a_i + k_1 a_{i+1})^2 + (k_1 b_{i-1} + k_2 b_i$$
$$+ k_1 b_{i+1})^2](k_1 b_{i-1} + k_2 b_i + k_1 b_{i+1}), i = 1 : n - 1 \tag{17}$$

$$\psi_j = (k_5 a_{i-1} + k_6 a_i + k_5 a_{i+1}) - q[(k_1 a_{i-1} + k_2 a_i + k_1 a_{i+1})^2 + (k_1 b_{i-1} + k_2 b_i$$
$$+ k_1 b_{i+1})^2](k_1 a_{i-1} + k_2 a_i + k_1 a_{i+1}), i = 1 : n - 1 \tag{18}$$

Finally, we solve interconnected systems of differential Eq. (16) (total $(2n + 2)$), simultaneously by SSP-RK54 method [23]. First we compute the initial solutions at $t = 0$ i.e a_i^0 and b_i^0 for $i = 0, 1, \ldots n$, using Thomas algorithm in the systems obtained in Sect. (3.1). After computing initial solutions, we solve the system (16) simultaneously using the Thomas algorithm in combination with the SSP-RK54 scheme.

4 Numerical Results

In this section, we apply the method for some problems discussed in the literature. It is well known that the analysis and dynamics of the nonlinear Schrödinger equation heavily rely on conservation principles. Many numerical approaches [8, 21, 24] have been developed in an effort to preserve Charge and Energy during simulations.

Therefore, we examine the applicability and accuracy of the scheme by evaluating the conservation laws.

One may demonstrate that an infinite number of conservation laws can be found for the pure initial-value problem (1–2) where the charge I_1 and the energy are the most prevalent of them [10], given by

$$I_1 = \int_{x_0}^{x_n} |F^{ex}|^2 dx \approx h \sum_{j=0}^{n} |F_j^{app}|^2,$$

$$I_2 = \int_{x_0}^{x_n} (|F_x^{ex}|^2 - \frac{q}{2}|F^{ex}|^4)dx \approx h \sum_{j=0}^{n} (|(F_x^{app})_j|^2 - \frac{q}{2}|F_j^{app}|^4),$$

where F^{ex} and F^{app} denotes the exact and approximate solutions, respectively. The relative changes of conserved quantities are given as

$$E - I_1 = \frac{I_1 - I_1^0}{I_1^0}, \quad E - I_2 = \frac{I_2 - I_2^0}{I_2^0}$$

where I_1^0 and I_2^0 are the quantities preserved over time $t = 0$ and I_1 and I_2 denotes the conserved quantities at time $t = t_k$. I_1 and I_2 are computed by using trapezoidal rule of integration.

4.1 Numerical Study of Single Soliton

First, we study the single soliton with the following analytic solution:

$$F(x, t) = f + ig = \gamma\sqrt{2/q}\,e^{i\left(\frac{1}{2}vx - \frac{1}{4}(v^2 - \gamma^2)t\right)} \operatorname{sech} \gamma(x - vt)$$

where v is speed of soliton and γ is the amplitude. Simulations have been performed with velocity of wave $v = 4$, $q = 2$ and amplitudes $\gamma = 1, 2$ in the domain $-20 \leq x \leq 20$.

The envelope soliton for amplitude $\gamma = 1$ is given by $|F| = \operatorname{sech}(x - vt)$. Simulation of single soliton with real and imaginary parts with amplitude $\gamma = 1$ has been shown in Fig. 1. It shows that, for the time span $0 \leq t \leq 4$, the soliton travels in the right direction at a constant speed of 4. Table 2 presents an analysis of the relative changes of invariants I_1 and I_2. The accuracy of the scheme at time $t = 2.5$ is demonstrated by comparing the recorded invariants with previous findings and listing them in Table 3. Tables 2 and 3 show that, in comparison to the other approaches [11, 14, 15], the current method is more dependable and preserves the relative change in quantities more effectively. Finally in Table 4 invariants are reported for all time $0 \leq t \leq 4$, which confirms the authenticity of obtained results. Further, the exact envelope soliton for amplitude $\gamma = 2$ is $|F| = \operatorname{sech} 2(x - vt)$.

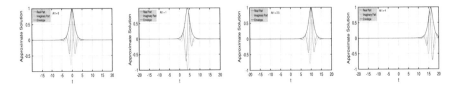

Fig. 1 Simulations of single soliton

Table 2 Invariants and their relative changes for single soliton at time $t = 1$ with amplitude $\gamma = 1$

Method	h	Δt	I_1	I_2	$E - I_1$	$E - I_2$
Proposed Scheme	0.09	0.002	1.99999	7.33354	–3.3339E-12	5.7857E-5
CDQ [14]	0.3125	0.025	1.999	7.333	–4.380E-6	–1.2143E-5
PDQ [15]	0.3125	0.020	–	–	–1.4403E-6	–3.9419E-6
A-L G. [24]	0.05	0.04	–	–	0.00003	0.00550
A-L L. [24]	0.06	0.0164	–	–	0.00004	–0.00797
Pseudospectral [24]	0.3125	0.00026	–	–	0.00001	–0.00003
Split-step [24]	0.3125	0.020	–	–	0.00000	0.00005
Hopscotch [24]	0.08	0.002	–	–	0.00003	–0.01407
Collocation [11]	0.05	0.005	–	–	0.00000	0.00000
Galerkin [6]	0.05	0.005	–	–	0.0000000	0.0000006
	0.3125	0.020	–	–	0.0000066	–0.0003417
Explicit scheme [24]	0.05	0.000625	–	–	0.00000	0.00556
Implicit/explicit scheme [24]	0.05	0.001	–	–	–0.00393	–0.01205
C-N [24]	0.05	0.005	–	–	–0.00001	–0.00557

Table 3 Invariants and their relative changes at time $t = 2.5$ with amplitude $\gamma = 1$

Method	h	Δt	I_1	I_2	$E - I_1$	$E - I_2$
Proposed Scheme	0.09	0.002	1.999999	7.333842	–8.51518E-12	9.84015E-5
FDM-DQM [3]	0.14	0.005	2.000008	7.333730	–	–
G [9]	0.3125	0.001	1.999908	7.333177	–	–
MQ [9]	0.3125	0.001	1.999472	7.331960	–	–
IMQ [9]	0.3125	0.001	1.999137	7.329795	–	–
IQ [9]	0.3125	0.001	1.999812	7.329801	–	–

To corroborate it numerically, a comparison of invariants is shown in Table 5 at $t = 1$ with earlier works [6, 11, 20, 24] and found good agreement with them. Graphs for time interval $0 \leq t \leq 4$ are also reported in Fig. 2. It is evident that the soliton has the largest amplitude $\gamma = 2$ and moves in the correct direction $\forall$ times t.

Table 4 Invariants and their relative changes at different times by proposed scheme with amplitude $\gamma = 1$

t	I_1	I_2	$E - I_1$	$E - I_2$
0	1.99999999	7.3331211	–	–
0.5	1.99999999	7.3333113	−1.681210E-12	2.5936E-5
1.5	1.9999999	7.3337193	−5.013101E-12	8.1573E-5
2.0	1.9999999	7.3338189	−6.731726E-12	9.5164E-5
3.0	1.9999999	7.3338007	−1.035871E-11	9.2672E-5
4.0	1.9999999	7.3318998	−1.420596E-11	1.6654E-4

Table 5 Invariants and their relative changes for single soliton at time $t = 1$ with amplitude $\gamma = 2$

Method	h	Δt	I_1	I_2	$E - I_1$	$E - I_2$
Proposed Scheme	0.09	0.002	3.99999	10.6714557	5.14981E-11	7.28541E-5
Quartic spline [6]	0.1	0.005	–	–	0.00000001	−0.000008
	0.1563	0.0048	–		0.0000095	−0.000276
Quintic Collocation [20]	0.015	0.005	–	–	0.0000000	0.0000001
	0.1	0.005	–	–	0.0000000	0.0000000
	0.1563	0.0048	–	–	0.0000000	0.0000026
	0.02	0.0025	–	–	0.0000000	0.0000000
A-L G. [24]	0.07	0.012	–	–	-0.00004	−0.03324
A-L L. [24]	0.06	0.03	–	–	-0.00001	−0.02526
Pseudospectral [24]	0.1563	0.0011	–	–	0.00000	0.00005
Split-step [24]	0.1563	0.0048	–	–	0.00000	0.00034
Hopscotch [24]	0.02	0.0004	–	–	0.00002	−0.00284
Collocation [11]	0.015	0.005	–	–	0.00000	0.00025
	0.02	.0025	–	–	0.00000	0.00004
Explicit [24]	0.02	0.0001	–	–	−0.00437	−0.00284
Implicit-explicit [24]	0.03	0.00022	–	–	0.00003	−0.02243
C-N [24]	0.02	0.011	–	–	0.00000	−0.00273

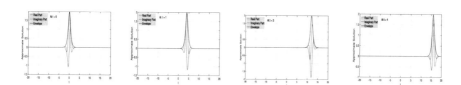

Fig. 2 Simulations of single soliton

4.2 Numerical Study of Double Soliton

Now, we study the interaction of two solitons with the following initial condition as given in [6, 11, 20]:

$$F(x, 0) = \sum_{j=1}^{2} F_j(x, 0)$$

where $F_j(x, 0) = \gamma_j \sqrt{2/q} e^{i\{\frac{1}{2}v_j(x-x_j)\}}$ sech $\gamma_j(x - x_j)$,. The parameters are taken as $q = 2$, $\gamma_1 = 1 = \gamma_2 = 1$, $v_1 = 4.0$, $v_2 = -4.0$, $x_1 = -10$, $x_2 = 10$.

The initial condition exhibits that there are two solitons of each of equal magnitude $\gamma_1 = \gamma_2 = 1$ and velocity equal $v_1 = v_2 = 4$ and initially at $t = 0$, their peak points are located at point $x_1 = -10$ and $x_2 = 10$. Simulation results depict that as time pass on soliton starts moving in opposite directions and comes closer. At time $t = 2$ they collide and after some time at $t = 2.5$ complete interaction take place, where amplitude is observed as $\gamma = 2$. After $t = 2.5$ the solitons again getting separated from each other see Fig. 3 and completely separate at time $t = 3.5$ again move in the opposite directions. The simulation is displayed in Fig. 3 and runs until time $t = 6$. The solitons are seen to travel in opposite directions, collide, and then split off while maintaining their initial characteristics, such as velocity, amplitude, and position.

Up to time $t = 6$, the preserved values and their relative changes are documented in Table 6. One can be noticed that throughout the simulation I_1, I_2 and their relative changes are found with a good degree of accuracy. It is observed that I_1 and I_2 remain the same to five digits. Further the affectivity of the scheme is tested by comparing the invariant with some published work. As can be shown, the invariants are accurately conserved with respect to the results obtained by the authors [6, 11, 24]. The comparison is shown at time $t = 1$ in Table 7 and at $t = 2.5$ in Table 8.

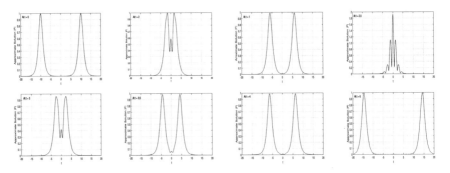

Fig. 3 Simulations of collison of two solitons

Table 6 Relative changes of invariants using $h = 0.1$, $\Delta t = 0.002$

t	$E - I_1$	$E - I_2$
0	–	–
1.0	–3.41271E-12	7.02267E-5
2.0	–3.63942E-12	7.78426E-5
2.5	–2.96872E-11	2.82579E-3
3.0	–3.59623E-11	1.24405E-4
3.5	–3.87679E-11	1.15417E-4
4.0	–4.16905E-11	9.35158E-5
6.0	–5.27018E-11	4.36287E-5

Table 7 Relative changes of invariants using $h = .1$, $\Delta t = .002$

	Proposed method		Gardner et al. [11]		Dag [6]	
t	I_1	I_2	I_1	I_2	I_1	I_2
0	3.999999991047675	14.666150205664573	3.99998	14.66596	3.99999	14.83143
1.0	3.999999991034024	14.667180161182770	3.99998	14.66706	3.99999	14.83157
2.0	3.999999991033118	14.667291858245640	3.99999	14.66693	3.99999	14.83261
2.5	3.999999990928926	14.624706679453990	3.99998	14.61440	3.99999	14.95380
3.0	3.999999990903826	14.667974762053991	3.99998	14.66789	3.99999	–
3.5	3.999999990892603	14.667842929573602	3.99999	14.66781	3.99999	14.83161
4.0	3.999999990880913	14.667521722974707	3.99998	14.66746	3.99999	14.83158
6.0	3.999999990836868	14.666790071074235	–	–	–	–

Table 8 Relative changes of invariants for collision of solitons at time $t = 2.5$

Method	h	Δt	$E - I_1$	$E - I_2$
Present method	0.1	0.002	–2.96872E-11	2.82579E-3
Implicit/explicit [24]	0.05	0.0025	0.00314	0.01434
Collocation [11]	0.1	0.01	0.00000	0.00120
C-N [24]	0.13	0.04	–0.00009	0.00619
Pseudospectral [24]	0.625	0.0071	0.00073	0.03247
Split-step [24]	0.625	0.005	0.00071	0.03595
Hopscotch [24]	0.05	0.001	0.00003	0.00063
Galerkin [6]	0.10	0.01	–0.000003	0.00330
Explicit [24]	0.13	0.0036	0.00000	0.00659
A-L G. [24]	0.08	0.045	–0.00012	0.00148
A-L L. [24]	0.07	0.07	0.00000	0.00156

4.3 Numerical Study of Birth of Soliton Using Maxwellian Initial Condition

The literature discusses how, given the initial condition $F(x, 0)$, a soliton should develop over time with a value greater than π if $I = \int_{-\infty}^{\infty} F(x, 0) \geq \pi$. If not, the soliton will decay away. Authors have used different initial conditions to study the birth of solitons. While Gardner et al. [11] found the identical simulation for a Maxwellian initial condition, denoted by $F(x, 0) = Ae^{-x^2}$, Delfour et al. [8] utilized a square well initial condition. This condition allows for the computation of analytical conserved quantities, which are as follows:

$$I_1 = A^2 \frac{\sqrt{\pi}}{2} \quad I_2 = \frac{A^2}{4} (2\sqrt{2} - q A^2) \sqrt{\pi}$$

We have also used the above Maxwellian initial condition in our simulation. Under Maxwellian conditions, $I = A\sqrt{\pi}$, soliton will develop if $A > \sqrt{\pi} = 1.7725$. Simulations are observed with parameter value $h = .1$, $\Delta t = .002$ up to time $t = 6$ in Fig. 4. For $A = 1$, the approximate solution $|F|$ decays with increasing time, but for $A = 1.78$, the amplitude, shape, and speed of the soliton remain unchanged. Fig. 4, which plots the maximum of solitons for time up to $t = 6$, further demonstrates that a soliton of magnitude 2 is formed if $A = 1.78 > \sqrt{(\pi)} = 1.7725$, and that it decays away if $A < \sqrt{(\pi)}$. To ensure the stability in the sense of conservation laws, we have computed the relative changes of invariants numerically for time up to $t = 6$ in Table 9 and comparison of approximate value I_1, I_2 has been also depicted in Table 10 with some earlier works. It is obvious that invariants are nearly constant throughout the simulation, which ensures the stability of the scheme in terms of conservation laws. It is also noted that conserved quantities obtained numerically and analytically preserved satisfactory well.

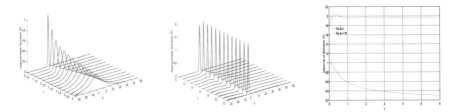

Fig. 4 Left: Simulation of decay of soliton for A = 1, Middle: Simulation of birth of soliton for A = 1.78 . Right: Maxima of solution for $0 \leq t \leq 6$

Table 9 Relative changes of invariants-$A = 1$ and $A = 1.78$ with $h = 0.1$, $\Delta t = 0.002$

A = 1			A = 1.78		
t	$E - I_1$	$E - I_2$	t	$E - I_1$	$E - I_2$
0	–	–	0	–	–
1	−4.04647E-13	4.86199E-3	1	2.01512E-12	9.67452E-4
2	−1.07522E-13	4.37173E-3	2	3.83430E-12	1.07852E-3
3	−1.75146E-12	4.06672E-3	3	5.47176E-12	1.05116E-3
4	−2.42451E-12	3.92973E-3	4	7.02211E-12	4.19754E-3
5	−3.09668E-12	3.48511E-3	5	8.72700E-12	4.24615E-3
6	−3.77150E-12	2.37354E-3	6	1.04022E-11	4.08212E-3

Table 10 Relative changes of invariants-$A = 1$ and $A = 1.78$ with $h = .1$, $\Delta t = .002$

	Proposed method		A.Bashan [3]		Saka [20]	
t	I_1	I_2	I_1	I_2	I_1	I_2
A = 1						
0	1.253314137315495	0.367051019296387	1.25331	0.36711	–	–
1	1.253314137314988	0.368835621347719	1.25331	0.36712	–	–
2	1.253314137314148	0.368655670685567	1.25331	0.36712	–	–
3	1.253314137313300	0.368543714847462	1.25330	0.36712	–	–
4	1.253314137312457	0.368493431380384	1.25330	0.36713	–	–
5	1.253314137311614	0.368330234909873	1.25330	0.36714	–	–
6	1.253314137310769	0.367922230972180	1.25329	0.36713	–	–
A = 1.78						
0	3.971000512670416	−4.925732293848144	3.97100	−4.92558	3.97100	−4.92562
1	3.971000512678420	−4.930389991978744	3.97098	−4.92566	3.97100	−4.93240
2	3.971000512685644	−4.930937094158369	3.97093	−4.92539	3.97100	−4.93377
3	3.971000512692147	−4.930802329584360	3.97085	−4.92496	3.97100	−4.93326
4	3.971000512698301	−4.946408289676421	3.97080	−4.92446	3.97100	−4.93335
5	3.971000512705071	−4.946647714564471	3.97074	−4.92385	3.97100	−4.93346
6	3.971000512711723	−4.945839752647428	3.97070	−4.92271	3.97100	−4.93298

4.4 Numerical Study of Birth of Mobile Soliton

With the initial condition $u(x, 0) = Ae^{-x^2+2ix}$, we have explored the formation of mobile solitons in the domain $-30 \le x \le 60$ in this experiment. The simulation is performed with the parameters $h = 0.1$, $\Delta t = 0.002$, for $A = 1$ and 1.78, over the domain $-30 \le x \le 60$ until time $t = 6$. We found that, for $A = 1$, the approximate solution $|f|$ decays with increasing time, and that, for $A = 1.78$, the soliton's amplitude, shape, and speed remain unchanged (Fig. 5). The soliton maxima shown in Fig. 5 further support the idea that a soliton of magnitude 2 is generated if $A = 1.78 > \sqrt{(\pi)} = 1.7725$, and that it decays away if $A < \sqrt{(\pi)} = 1$. The invariant and relative changes of invariants for time up to $t = 6$ for $A = 1$ and $A = 1.78$ is

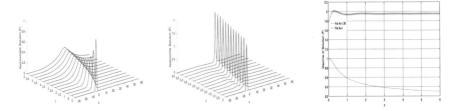

Fig. 5 Left: Simulations of decay away of soliton for A = 1, Middle: Simulations of birth of soliton for A = 1.78. Right: Maxima of approximate solution for $0 \leq t \leq 6$

shown in Tables 11 and 12, respectively, and also compared with some earlier works. Results show that the invariants are nearly constant throughout the simulation and captured reasonably very well in comparison with some earlier work. Hence the stability of the scheme is again guaranteed in terms of conservation laws.

4.5 Numerical Study of Bound State of Soliton

To study the bound state of solitons, Miles [17] has obtained accurate results by using the following condition

$$u(x, 0) = \text{sech}(x) \tag{19}$$

It yields a bound state of λ solutions, if $q = 2\lambda^2$, $\lambda = 1, 2$ and the solution lacks the usable form if $\lambda \geq 3$. The values of conservation are calculated using:

$$I_1 = 2, \; I_2 = \frac{2}{3}(1 - q) \tag{20}$$

Numerical results have been obtained for bound state of solitons with $h = .08$, $\Delta t = .001$ for $q = 18$ and 32 over a region $-20 \leq x \leq 20$. In Table 13, conservation of conserved quantities and their relatives changes have been shown numerically. It is evident that both invariants stay relatively constant. I_1 and I_2 have converged to seven and three digits, respectively, up to time $t = 5$. To visualize the solution behavior clearly for $\lambda = 3$, figures have been plotted separately for time $0 \leq t \leq 1$ and again for $0 \leq t \leq 5$. We can easily see through Fig. 6 that for starting times, the initial solitons gradually transform, becoming narrower and taller, developing two peaks on either side, and eventually return to their original form around $t = 0.2$. Following that, it undergoes a split down the middle, resulting in two peaks on either side. Then, around t=0.5, it returns to its initial shape. The soliton once more undergoes a transformation, becoming narrower and taller. At roughly t=0.75, its shape finally fully returns to its initial state. This periodic behavior continues to

Table 11 Relative changes of invariants for $A = 1$ with $h = 0.1$, $\Delta t = 0.002$

t	Proposed method				A.Bashan [3]			
	I_1	I_2	$E - I_1$	$E - I_2$	I_1	I_2	$E - I_1$	$E - I_2$
0	1.2533141437315495	5.378549101674104			1.25331	5.38148		
1	1.2533141437151341	5.402140931863491	−1.30976E-10	4.38628E-3	1.25324	5.37853	-5.9E-5	-5.5E-4
2	1.2533141437009708	5.406253966504101	−2.43983E-10	5.15099E-3	1.25324	5.37795	-6.1E-5	-6.6E-4
3	1.2533141436875793	5.407749636967544	−3.50832E-10	5.42907E-3	1.25323	5.37768	-6.5E-5	-7.1E-4
4	1.2533141436744797	5.408530575670008	−4.55351E-10	5.57426E-3	1.25324	5.37761	-6.2E-5	-7.2E-4
5	1.2533141436615061	5.409013119424407	−5.58865E-10	5.66398E-3	1.25325	5.37758	-5.5E-5	-7.3E-4
6	1.2533141436486032	5.409339616930978	−6.61816E-10	5.72468E-3	1.25328	5.37752	-2.9E-5	-7.4E-4

Table 12 Relative changes of invariants for $A = 1.78$ with $h = .1$, $\Delta t = .002$

	Proposed method				A.Bashan [3]		Saka [20]	
t	I_1	I_2	$E - I_1$	$E - I_2$	I_1	I_2	I_1	I_2
0	3.971000512670417	10.952698230357406			3.97100	10.96012	3.97100	10.95837
1	3.971000510676921	10.912897184359498	-5.02013E-10	-3.63390E-3	3.97111	10.96130	3.97100	10.97104
2	3.971000509359589	10.903962000342499	-8.33751E-9	-4.44970E-3	3.97114	10.96184	3.97100	10.97294
3	3.971000508108381	10.902002930662839	-1.14883E-9	-4.62856E-3	3.97114	10.96234	3.97100	10.97289
4	3.971000506926736	10.908242120191220	-1.44640E-9	-4.05891E-3	3.97114	10.96272	3.97100	10.97336
5	3.971000505633611	10.901557474862468	-1.77204E-9	-4.66923E-3	3.97112	10.96289	3.97100	10.97374
6	3.971000504441602	10.905953010444280	-2.07222E-9	-4.26791E-3	3.97107	10.96299	3.97100	10.97592

Table 13 Invariants and their relative changes with $h = .08$, $\Delta t = .001$ for $q = 18$

t	I_1	I_2	$E - I_1$	$E - I_2$
0	1.99999999	−11.3333337	–	–
1.5	1.99999995	−11.3436624	−2.4174E-8	9.1135E-4
3.0	1.99999990	−11.3365715	−4.8060E-8	2.8568E-4
4.5	1.99999982	−11.3783225	−8.8061E-8	3.9695E-3
5.0	1.99999981	−11.3581071	−9.3442E-8	2.1858E-3

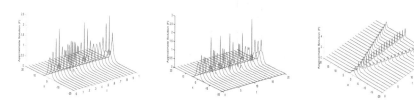

Fig. 6 Left: Simulation of soliton for $\lambda = 3$ and $0 \leq t \leq 1$. Middle: Simulation of soliton for $\lambda = 3$ and $0 \leq t \leq 5$. Right: Simulation of soliton for $\lambda = 4$ and $0 \leq t \leq 5$

be observed for extended time intervals within the range $0 \leq t \leq 5$ and depicted in Fig. 6. Figure 6 also displays the graphs for $\lambda = 4$. The behavior and figures of the three- and four-bound solitons are in good agreement with the papers [14, 15, 18].

5 Conclusions

In the present study, we offer a numerical solution for the one-dimensional Schrödinger equation. The modified form of trigonometric cubic B-spline functions provides an efficient way to deal with the Dirichlet boundary conditions. The computational complexity of the scheme is observed in terms of a number of simple arithmetic operations used and found to be $O(n)$, n is a number of nodes. The current approach's strength is that it doesn't require complicated calculations and is easy to implement. Numerical simulations have effectively explained the various physical phenomena and validated charge and energy conservation rules. The proposed approach is novel to find solutions of nonlinear Schrödinger equation to the best knowledge of the authors. Thus, it is suggested for future research work that this method can be developed further to find numerical solutions to higher dimensional differential equations.

References

1. Murat Aksoy A, Dag DI (2013) Taylor-collocation method for the numerical solution of the nonlinear schrodinger equation using cubic b-spline basis. Int J Nonlinear Sci 15(4):322–333
2. Antoine X, Besse C, Descombes S (2006) Artificial boundary conditions for one-dimensional cubic nonlinear schrödinger equations. SIAM J Num Anal 43(6):2272–2293
3. Bashan A (2019) A mixed methods approach to schrodinger equation: Finite difference method and quartic b-spline based differential quadrature method. Int J Optim Control Theories Appl (IJOCTA) 9(2)
4. Borhanifar A, Abazari R (2010) Numerical study of nonlinear schrödinger and coupled schrödinger equations by differential transformation method. Opt Commun 283(10):2026–2031
5. Braun M (2023) Numerical solution of the one dimensional schrödinger equation using a basis set of scaled and shifted sinc functions on a finite interval. J Comput Appl Math 429:115224
6. Dag I (1999) A quadratic b-spline finite element method for solving nonlinear schrödinger equation. Comput Methods Appl Mech Eng 174(1):247–258
7. Dehghan M (2006) Finite difference procedures for solving a problem arising in modeling and design of certain optoelectronic devices. Math Comput Simul 71(1):16–30
8. Delfour M, Fortin M, Payr G (1981) Finite-difference solutions of a non-linear schödinger equation. J Comput Phys 44(2):277–288
9. Dereli Y, Irk D, Dag I (2009) Soliton solutions for nls equation using radial basis functions. Chaos Solitons Fractals 42(2):1227–1233
10. Fei Z, Pérez-García VM, Vázquez L (1995) Numerical simulation of nonlinear schrödinger systems: A new conservative scheme. Appl Math Comput 71(2):165–177
11. Gardner LRT, Gardner GA, Zaki SI, El Sahrawi Z (1993) B-spline finite element studies of the nonlinear Schrödinger equation. Comput. Methods Appl Mech Eng 108(3–4):303–318
12. Hong J, Ji L, Liu Z (2018) Optimal error estimate of conservative local discontinuous galerkin method for nonlinear schrödinger equation. Appl Numer Mathe 127:164–178
13. Iqbal A, Abd Hamid NN, Md. Ismail AI (2020) Cubic b-spline galerkin method for numerical solution of the coupled nonlinear schrödinger equation. Math Comput Simul 174:32–44
14. Korkmarz A, Dağ İ (2008) A differential quadrature algorithm for simulations of non-linear schröninger equation. Comput Math Appl 56(9):2222–2234
15. Korkmaz A, Dağ İ (2008) A differential quadrature algorithm for nonlinear schrödinger equation. Nonlinear Dyn 56(1):69–83
16. Lu W, Huang Y, Liu H (2015) Mass preserving discontinuous galerkin methods for schrödinger equations. J Comput Phys 282:210–226
17. Miles JW (1981) An envelope soliton problem. SIAM J Appl Math 41(2):227–230
18. Mokhtari R, Isvand D, Chegini NG, Salaripanah A (2013) Numerical solution of the schrödinger equations by using delta-shaped basis functions. Nonlinear Dyn 74(1):77–93
19. Rani R, Arora G, Emadifar H, Khademi M (2023) Numerical simulation of one-dimensional nonlinear schrodinger equation using pso with exponential b-spline. Alexandria Eng J 79:644–651. https://doi.org/10.1016/j.aej.2023.08.050
20. Saka B (2012) A quintic b-spline finite-element method for solving the nonlinear schrödinger equation. Phys Wave Phenom 20(2):107–117
21. Sanz-Serna J, Manoranjan V (1983) A method for the integration in time of certain partial differential equations. J Comput Phys 52(2):273–289
22. Sheng Q, Khaliq A, Al-Said E (2001) Solving the generalized nonlinear schrödinger equation via quartic spline approximation. J Comput Phys 166(2):400–417
23. Spiteri RJ, Ruuth SJ (2002) A new class of optimal high-order strong-stability-preserving time discretization methods. SIAM J Numer Anal 40(2):469–491 (Electronic). https://doi.org/10.1137/S0036142901389025
24. Taha TR, Ablowitz MJ (1984) Analytical and numerical aspects of certain nonlinear evolution equations. ii. numerical, nonlinear schrödinger equations. J Comput Phys 55:203–230

25. Zakharov VE, Shabat AB (1971) Exact theory of two-dimensional self-focusing and one-dimensional self-modulation of waves in nonlinear media. Ž. Èksper. Teoret Fiz 61(1):118–134
26. Zhou S, Cheng X (2010) Numerical solution to coupled nonlinear schrödinger equations on unbounded domains. Math Comput Simul 80(12):2362–2373

Fuzzy MCDM Techniques for Analysing the Risk Factors of COVID-19 and FLU

M. Sheela Rani and S. Dhanasekar

Abstract Decision-making is inevitable in day-to-day life. Fuzzy multi-criteria decision-making is incorporated for better decision-making in almost all kinds of complex problems. In this research notable advantages of VIKOR and TOPSIS techniques employed with fuzzy triangular numbers to analyse the risk factors of COVID-19 and FLU. The comparative analysis is illustrated to find the most influencing risk factors of COVID-19 and FLU by comparing each and every situation of patients. At last, the resistance test was also included to check the final rankings and outcome.

Keywords MCDM · Fuzzy set · COVID-19 and FLU · VIKOR · TOPSIS

1 Introduction

World Health Organisation (WHO) [1] provided a report that the COVID-19 epidemic has caused huge energy, economic, and social problems from the year 2019. In abroad countries, most of the research work is going on only for medical diagnosis using several mathematical tools. By this literature review, we can estimate that MCDM is also a part of medical diagnosis and it is doing a vital role. Sotoudeh et al. [2] explained that the COVID-19 pandemic has stimulated a panic of an extensive economic dilemma, destroyed over and above five million people widespread and had a wrinkle result in every situation of life. Hung Nguyen et al. [3] conducted a case study in Vietnam. They explained the situation in Vietnam at the time of the COVID-19 era. Alamoodi et al. [4] found that MCDM can aid in the development of every appropriate judgement group, elimination scheme, and vaccination evolution

M. Sheela Rani · S. Dhanasekar (✉)
Mathematics, School of Advanced Sciences, Vellore Institute of Technology, Chennai 600127, Tamilnadu, India
e-mail: dhanasekar.sundaram@vit.ac.in

M. Sheela Rani
e-mail: sheelarani.m2020@vitstudent.ac.in

© The Author(s), under exclusive license to Springer Nature Singapore Pte Ltd. 2024
D. Giri et al. (eds.), *Proceedings of the Tenth International Conference on Mathematics and Computing*, Lecture Notes in Networks and Systems 963,
https://doi.org/10.1007/978-981-97-2069-9_12

159

with a particular outcome. Educational literature demonstrates that MCDM is used at the epidemic in different samples and events, especially with reference to essential learning or technical augment. Chowdhury et al. [5] did research on cough sounds who are all affected by COVID-19 and they utilised machine learning techniques to find a disease level. Wise et al. [6] explained that vaccinating against COVID-19 and flu at the same time is safe in the UK. Jones et al. [7] did research on how the SARS-COVID-19 virus changes the cell structure of one own cell to another structure of the virus. Kiseleva et al. [8] compiled important biological features of naturally various microbes such as SARS-CoV-2, influenza viruses, and rhinoviruses, in a step they initiated their differences and similitude. Jaklevic et al. [9] explained about the vaccination during the health suffering from flu. Gasmi et al. [10] explained that mostly half of the patients described have a minimum of one symptom of COVID-19 by admission to the hospital. Diabetes, hypertension, chronic liver disease, cardio-vascular diseases, and obesity are the most generally reported. Comorbidity means a patient having two or more diseases at the same time. Comorbidity contributes to finding disease levels and hazards of tough symptoms. More than 80% of patients who need ICU have been kept to have comorbidity. Ozsahin et al. [11] explained the strengths and weaknesses of the selected vaccines, and they found by using the techniques of fuzzy PROMETHEE. Ecer et al. [12] selected the vaccines for patients according to their age, and they explained the reason why they chose the intuition-istic fuzzy number. That is intuitionistic fuzzy sets consist of a membership and a non-membership function, uncertainty and vagueness can be managed more user friendly. Apart from that, intuitionistic fuzzy sets are used for reporting the obscure resolution details. Yildrim et al. [13] utilised fuzzy PROMETHEE and fuzzy VIKOR techniques to choose the patients for an ICU ward. Alsalem et al. [14] provided a result for researchers and an overview of the new situation of MCDM consideration, developing techniques and motivation in tacking MCDM possibilities in tackling an approximate conclusion for various areas against COVID-19 and FLU. Ali et al. [15] provided the report about the approximate position of water bodies and the most accountable alternative with more weightage amidst every chosen choices, proximity to waste storage bins, surface temperature, moisture content and described malaria and dengue situations also shared the high-level donations in the importance of more diseases. Jafar et al. [16] did a comment that each MCDM approach provides a unique rank from those of other methods. Yas et al. [17] did an evaluation preference of the diabetes symptoms for patients by adapting decision-making techniques, and they used the TOPSIS technique to find the worst patient level. Hence, in the following work, we discussed patients who have the symptoms of COVID-19 and FLU disease.

2 Literature Review

Hwang and yoon [18] introduced the TOPSIS technique for the various levels of difficult decision fields. This is one of the distance-based techniques. The positive ideal solution is referred to as the shortest distance, and the negative ideal solution is referred to as the farthest distance. The relative rankings of the shortest and farthest distances are calculated by using the Euclidean distance. Naeem et al. [19] described an algorithm by joining the techniques of TOPSIS, VIKOR via Pythagorean fuzzy set in a medical diagnosis field, they provided the results of symptoms of hepatitis A to hepatitis E and also they utilised the technique for the patient management system. Akram et al. [20] provided guidance to determine the patient's health level and evaluate the disease and symptoms in a bipolar fuzzy environment. They utilised TOPSIS and ELECTRE-I and compared the ranking values to find an accurate ranking. Mohammed et al. [21] evaluated the COVID-19 diagnostic model by using the benchmarking methodology. They proved the algorithm by using fifty samples of chest X-ray datasets.

3 Preliminaries

This preliminary section outlines key concepts, definitions, and relevant context, setting the stage for the subsequent discussion.

Definition 1 *(Fuzzy set)* The fuzzy set was introduced by Lotfi Zadeh [22] in 1965, and it is defined as a set whose elements have a degree of membership. If A is a universal set with element a, then a fuzzy set $\tilde{X}$ in A is a set of ordered pairs,

$$\tilde{X} = \{(a, \mu_X(a)) : a \in A\}.$$

where, $\mu_X(a)$ is a membership function. The membership function $\mu_X(a))$ is defined as $\mu_A(x) : A \rightarrow [0, 1]$.

Definition 2 *(Fuzzy number)* Fuzzy number [23] is defined as a fuzzy set whose membership function should satisfy the following conditions A fuzzy set $\tilde{A}$ defined on the set of real numbers $\mathbb{R}$ is said to be a fuzzy number if its membership function $\tilde{A} : \mathbb{R} \rightarrow [0, 1]$ has the following characteristics.

(i) $\tilde{A}$ is convex
  $\mu_{\tilde{A}}(\lambda x_1 + (1 - \lambda)x_2) \geq \min(\mu_{\tilde{A}}(x_1), \mu_{\tilde{A}}(x_2)), \forall x \in [0, 1], \lambda \in [0, 1]$
(ii) $\tilde{A}$ is normal, (i.e.) $\exists$ an $x \in \mathbb{R}$ such that if $\max \mu_{\tilde{A}}(x) = 1$.
(iii) $\tilde{A}$ is piecewise continuous.

Definition 3 *(Triangular membership function)* A triangular membership function [24] is defined by three parameters (a, b, c) with $(a < b < c)$. These three parameters determine the coordinates of three angles a, b, c and triangular fuzzy number is defined as, A triangular fuzzy number $\tilde{A}$ denoted by (a, b, c) and the membership function is defined as

$$\mu_{\tilde{A}}(x) = \begin{cases} \dfrac{x-a}{b-a}, & \text{for } a \le x \le b \\ \dfrac{c-x}{c-b}, & \text{for } b \le x \le c \\ 0, & \text{elsewhere.} \end{cases}$$

3.1 Arithmetic Operations

The arithmetic operations of the triangular fuzzy number [25] are defined as follows:
Let $\tilde{A} = (x_1, x_2, x_3)$ and $\tilde{B} = (y_1, y_2, y_3)$.

1. Addition: $\tilde{A} + \tilde{B} = \{(x_1 + y_1), (x_2 + y_2), (x_3 + y_3)\}$
2. Subtraction: $\tilde{A} - \tilde{B} = \{(x_1 - y_3), (x_2 - y_2), (x_3 - y_1)\}$
3. Multiplication: $\tilde{A} * \tilde{B} = \{(x_1 * y_1), (x_2 * y_2), (x_3 * y_3)\}$
4. Division: $\tilde{A}/\tilde{B} = \{(x_1/y_3), (x_2/y_2), (x_3/y_1)\}$

4 Analysing the Risk Factors of COVID-19 and FLU

This section focuses on assessing the risk factors associated with both COVID-19 and influenza. This section delves into a comparative examination of the shared and distinct elements influencing the spread, severity, and vulnerability of these two infectious diseases.

4.1 Risk Factors of COVID-19 and FLU

Fever is the common risk factor of every viral disease [26]. Nowadays fever is spreading through bacteria and fungi, and is spreading to people and animals in a facile way [27]. WHO reports [1] in the starting stage of COVID-19 more than 60% people were affected by the fever. **Cough** and throat pain [5] plays a vital role in each disease. Cough is a normal symptom of many more diseases which spread easily. **Tiredness** or fatigue is defined as such a normal pain in the human body [28]. In this work, tiredness is referred to as a normal agony that can be accepted by a human body in a normal temperature and situation. **Loss of taste or smell** is an important and severe risk factor for COVID-19 and FLU. It is an early-stage symptom of COVID-19. In this work loss of taste is referred to as the molecules that dissolve to activate the sense of taste or smell. It can be a recovery risk factor of COVID-19 compared to other symptoms and approximately it might take two weeks of time to recover the cells of taste or smell [29]. **Sore throat** is a risk factor of COVID-19 which is reported to be in 5% to 17.4% of COVID-19 patients [30]. During the pandemic sore throat is a fundamental symptom that should be informed to doctors.

Headache is a specific risk factor COVID-19, and absorbingly 50% of patients of COVID-19 had no headache previously[31]. Headache is a common symptom of more viral diseases such as malaria and dengue. In this work **aches or pain** is referred to as a minor pain that can be felt by a patient, which is different from a headache. **Diarrhoea** is a low level symptom of COVID-19 and FLU [32], which is increasing the severity level of disease. **Red or irritated eyes** can lead to [33] a nose pain and other vein issues. Irritated eyes are a normal level symptom of FLU in an early stage. **Difficulty in breathing** is a dangerous symptom of COVID-19. More than 90% of the patients died who had difficulty breathing and it is an important risk factor for latent stage patients [34]. **Loss of speech** is a rare symptom of COVID-19 and other viral diseases, which affected 0.01% of patients who affected by COVID-19 [35]. **Body or chest pain** is a different one which is not the same as aches and other pains like headaches. Hence, by the review of medical diagnosis articles we have chosen the risk factors of COVID-19 and FLU.

4.2   Stages of COVID-19 and FLU

Normal stage is the starting stage of COVID-19 and FLU. In this stage immune system starts to fight with viruses and bacteria. It continues with no symptoms or signs. This type of infection does not lead to death or any other dangerous situation. **Borderline COVID-19 and FLU** is the time that the virus entered and it is changing the immune system to a poor one. Patients can have normal symptoms like cough, headache, aches, and pains. **Early COVID-19 and FLU** stage patients can estimate the situation of disease and it may lead to a latent situation. At this time diagnosis may be difficult. Radiotherapy systems can find the virus in a patient's body. **Latent COVID-19 and FLU** is final stage of the disease. There are more symptoms like shortness of breath, high-level fever, and some other latent level symptoms. In this situation, the immune system can get the most level of damage by the virus, and it may lead to recovery of the immune system if it has a fine energy level. WHO reports that 85% of people who were affected by COVID-19 and FLU are at this latent stage. Hence, we have chosen these four stages to find the critical level of patients who are affected by both disease COVID-19 and FLU.

Algorithm 1 Choose the R_i from $R_1, R_2, \ldots R_n$

Require: $n \neq 0 \; \forall \; m \neq 0$
Ensure: $R = \{R_1, R_2, \ldots R_m\}$ and $C = \{C_1, C_2, \ldots C_n\}$
1: Collect the data from the experts $D = D_1, D_2, \ldots, D_k$.
2: Aggregate the triangular fuzzy matrix by CFCS algorithm from the experts' opinion

$$
\begin{bmatrix}
r_{11} & r_{12} & r_{13} & \cdots & r_{1n} \\
r_{21} & r_{22} & r_{23} & \cdots & r_{2n} \\
\vdots & \vdots & \vdots & \ddots & \vdots \\
r_{m1} & r_{m2} & r_{m3} & \cdots & r_{mn}
\end{bmatrix}
$$

3: Determination of normalised fuzzy matrix $n_{ij} = \dfrac{r_{ij}}{\sqrt{\Sigma_{i=1}^{n} r_{ij}^2}}$
4: Calculation of weighted normalised fuzzy matrix

$$
N = \begin{bmatrix}
w_1 n_{11} & w_2 n_{12} & w_3 n_{13} & \cdots & w_n n_{1n} \\
w_1 n_{21} & w_2 n_{22} & w_3 n_{23} & \cdots & w_n n_{2n} \\
\vdots & \vdots & \vdots & \ddots & \vdots \\
w_1 n_{m1} & w_2 n_{m2} & w_3 n_{m3} & \cdots & w_n n_{mn}
\end{bmatrix}
$$

5: Calculation of positive (V^+) ideal solution and negative (V^-) ideal solution

$$
V^+ = \{max(N_{ij}) | j \in R), min(N_{ij}) | j \in C\}, \quad V^- = \{min(N_{ij}) | j \in R), max(N_{ij}) | j \in C\}
$$

6: Calculation of distance measure through TOPSIS technique,

$$
S_i^+ = \sqrt{\sum_{j=1}^{k} (N_{ij} - N_j^+)^2}, \quad S_i^- = \sqrt{\sum_{j=1}^{k} (N_{ij} - N_j^-)^2}
$$

7: Determine the relative closeness

$$
C_i^* = \frac{S_i^-}{S_i^+ + S_i^-}
$$

8: Estimate of value of P_i and R_i.
9: Find the value of Q_j

$$
Q_j = V \frac{(P_j - P^*)}{(P^- - P^*)} + (1 - V) \frac{(R_j - R^*)}{(R^- - R^*)}
$$

10: Determine the ranking value by descending order and make comparison of TOPSIS and VIKOR.

By Table 1, readers can understand the risk factors of COVID-19 and FLU. These are the important risk factors which are explained by WHO [1].

The criteria of Table 3 is referred to as the critical stages of COVID-19 and FLU. Through these four stages, patients suffer in various ways.

Linguistic terms are explained in Table 2 for fuzzy triangular numbers. The linguistic terms are taken from the fuzzy number between 0 and 1. The first decision-maker opinion is explained as Table 4. It is given by Dr. Kavya, Sathya Sai Medical College And Research Institute, Chennai. Decision-makers opinion was given by Tables 4, 5 and 6 are indicated the values of living chance of patients at each level of criterion and risk factor. In their opinion, risk factors and criteria were joined by

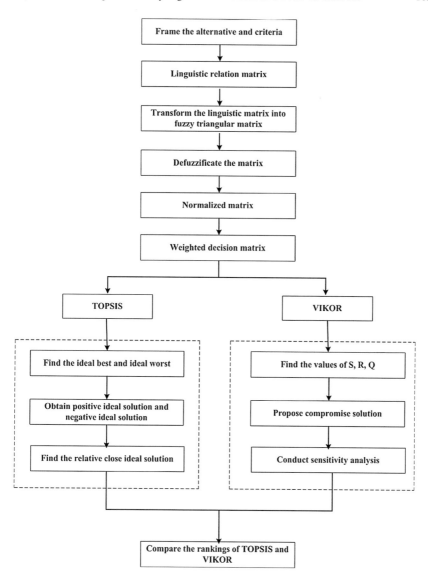

Fig. 1 Algorithm of hybrid technique

Table 1 Risk factors of COVID-19 and FLU

Alternatives	Notation
Fever	R_1
Cough	R_2
Tiredness	R_3
Loss of taste or smell	R_4
Sore throat	R_5
Headache	R_6
Aches or pain	R_7
Diarrhoea	R_8
Red or irritated eyes	R_9
Difficulty in breathing	R_{10}
Loss of speech	R_{11}
Body or chest pain	R_{12}

Table 2 Linguistic terms for triangular fuzzy sets

Linguistic terms	Fuzzy triangular numbers
Very Low (VL)	(0, 0, 0.25)
Low (L)	(0, 0.25, 0.5)
Medium (M)	(0.25, 0.50, 0.75)
High (H)	(0.50, 0.75, 1.00)
Very High (VH)	(0.75, 1.00, 1.00)

Table 3 Stages of COVID-19 and FLU

Criteria	Notation
Normal (Molecular)	C_1
Borderline COVID-19 and FLU(Biochemical cardiometabolic risk)	C_2
Early COVID-19 and FLU (Biochemical disease)	C_3
Late COVID-19 and FLU (Vascular complications)	C_4

utilising the fuzzy triangular number. Table 6 is given by the third decision-maker Dr. Rihan Khan, Sathya Sai Medical College And Research Institute, Chennai.

Table 7 denoted as the weights collected from the decision-maker about the stages of patients. In Table 8, the decision-makers opinions measured as a fuzzy triangular number and it is converted to aggregated fuzzy triangular matrix.

Table 9 is defuzzified form of the aggregated triangular fuzzy matrix.

Table 4 First decision-maker opinion about COVID-19 and FLU

Alternatives	C_1	C_2	C_3	C_4
R_1	VH	VH	H	M
R_2	VH	VH	H	M
R_3	VH	VH	H	M
R_4	VH	VH	H	M
R_5	VH	VH	H	M
R_6	VH	VH	H	M
R_7	VH	VH	H	M
R_8	VH	VH	H	M
R_9	VH	VH	H	M
R_{10}	VH	H	M	L
R_{11}	VH	H	M	M
R_{12}	VH	H	M	M

Table 5 Second decision-maker opinion about COVID-19 and FLU

Alternatives	C_1	C_2	C_3	C_4
R_1	H	H	M	M
R_2	VH	H	H	M
R_3	H	H	M	M
R_4	VH	VH	M	M
R_5	H	M	M	M
R_6	M	M	M	L
R_7	H	H	H	M
R_8	H	M	L	VL
R_9	H	H	H	M
R_{10}	M	H	L	M
R_{11}	VH	VH	H	M
R_{12}	M	M	M	M

Table 11 is the collection of the highest and lowest values of the weighted matrix, which can be from each stage of the disease.

Through Table 12, the values of S_i^+ and S_i^- collected.

Table 13 is the ranking values that we got from the TOPSIS technique.

Through Fig. 2a, readers can analyse the rankings and values of each risk factor. Table 14 is expressing the values of P_i and R_i which we got for the VIKOR technique. In this stage, choose the best and the worst values from the TOPSIS method Tables 11 and 15 denoted that, the values of P^*, R^*, P^- and R^- got from the Table 14.

In Table 16, readers can understand the values of Q_i to find the rankings in descending order.

Table 6 Third decision-maker opinion about COVID-19 and FLU

Alternatives	C_1	C_2	C_3	C_4
R_1	VH	H	H	M
R_2	VH	H	H	M
R_3	VH	H	M	M
R_4	VH	H	M	M
R_5	VH	H	H	M
R_6	VH	H	H	M
R_7	VH	H	H	M
R_8	VH	H	H	M
R_9	VH	H	H	M
R_{10}	VH	H	L	M
R_{11}	VH	H	H	M
R_{12}	VH	H	M	M

Table 7 Collective weights from decision-makers for the symptoms of FLU and COVID-19

Decision maker	C_1	C_2	C_3	C_4
DM_1	VH	H	M	L
DM_2	VH	VH	L	L
DM_3	VH	H	M	M

Table 8 Aggregated triangular fuzzy matrix

Risk factors	C_1	C_2	C_3	C_4
R_1	(0.50, 0.917, 1)	(0.50, 0.83, 1)	(0.25, 0.58, 1)	(0.25, 0.50, 0.75)
R_2	(0.75, 1, 1)	(0.50, 0.917,1)	(0.25, 0.67, 1)	(0.25, 0.50, 0.75)
R_3	(0.50, 0.917, 1)	(0.50, 0.917, 1)	(0.25, 0.67, 1)	(0.25, 0.50, 0.75)
R_4	(0.75,1,1)	(0.50, 0.92, 1)	(0.25, 0.67, 1)	(0.25, 0.50, 0.75)
R_5	(0.50, 0.917, 1)	(0.25, 0.75, 1)	(0.25, 0.58, 1)	(0.25, 0.50, 0.75)
R_6	(0.25, 0.67, 1)	(0.25, 0.75, 1)	(0.25, 0.58, 1)	(0, 0.42, 0.75)
R_7	(0.50, 0.917, 1)	(0.50, 0.917, 1)	(0.25, 0.58, 1)	(0.25, 0.50, 0.75)
R_8	(0.50, 0.917, 1)	(0.25, 0.75, 1)	(0, 0.58, 1)	(0, 0.33, 0.75)
R_9	(0.50, 0.917, 1)	(0.50, 0.83, 1)	(0.50,0.75,1.00)	(0.25, 0.50, 0.75)
R_{10}	(0.25, 0.67, 1)	(0.25, 0.67, 1)	(0.25, 0.67, 1)	(0, 0.42, 0.75)
R_{11}	(0.75,1.00,1.00)	(0.50, 0.67, 1)	(0.25, 0.67, 1)	(0.25, 0.50, 0.75)
R_{12}	(0.25, 0.67, 1)	(0.25, 0.67, 1)	(0.25, 0.58, 1)	(0.25, 0.50, 0.75)

Table 9 Defuzzified matrix

Risk factors	C_1	C_2	C_3	C_4
R_1	0.888	0.803	0.585	0.488
R_2	0.962	0.888	0.637	0.488
R_3	0.888	0.888	0.637	0.488
R_4	0.962	0.888	0.637	0.488
R_5	0.888	0.696	0.585	0.488
R_6	0.65	0.696	0.585	0.400
R_7	0.888	0.888	0.585	0.488
R_8	0.888	0.696	0.546	0.349
R_9	0.888	0.803	0.733	0.488
R_{10}	0.650	0.650	0.637	0.400
R_{11}	0.962	0.692	0.637	0.488
R_{12}	0.650	0.650	0.585	0.488

Table 10 Normalised matrix

Risk factors	C_1	C_2	C_3	C_4
R_1	0.3	0.299	0.273	0.304
R_2	0.325	0.33	0.298	0.304
R_3	0.3	0.33	0.298	0.304
R_4	0.325	0.33	0.298	0.304
R_5	0.3	0.259	0.273	0.304
R_6	0.219	0.259	0.273	0.249
R_7	0.3	0.33	0.273	0.304
R_8	0.3	0.259	0.255	0.217
R_9	0.3	0.299	0.343	0.304
R_{10}	0.219	0.242	0.298	0.249
R_{11}	0.325	0.257	0.298	0.304
R_{12}	0.219	0.242	0.273	0.304

Table 11 Selecting the best and worst values

Selected values	C_1	C_2	C_3	C_4
Best value (V^+)	0.146	0.0825	0.05145	0.0456
Worst value (V^-)	0.0986	0.0605	0.03825	0.03255

Table 12 Finding the value of S_i^+ and S_i^-

Risk factors	S_i^+	S_i^-
R_1	0.01703	0.04136
R_2	0.00675	0.05422
R_3	0.01287	0.04483
R_4	0.00675	0.0542
R_5	0.02345	0.03899
R_6	0.05244	0.07071
R_7	0.01183	0.04461
R_8	0.02145	0.03647
R_9	0.01342	0.03987
R_{10}	0.05339	0.07071
R_{11}	0.01949	0.0493
R_{12}	0.05329	0.07071

Table 13 Ranking values of TOPSIS technique

Risk factors	C_i^*	Rank
R_1	1.41175	7
R_2	1.12412	12
R_3	1.28708	10
R_4	1.12454	11
R_5	1.61701	3
R_6	1.48088	5
R_7	1.48083	6
R_8	1.58815	4
R_9	1.33659	9
R_{10}	1.75506	1
R_{11}	1.39533	8
R_{12}	1.75364	2

5   Discussion

As a way of approving the hybrid method, this work was designed through the distance-based techniques TOPSIS and VIKOR. In early stage, the respected risk factors and stages of patients based on the decision maker's opinion. Through Table 17 readers can analyse the risk factors and its rankings in the disease of COVID-19 and FLU. By the analysis of these hybrid techniques, the risk factor R_{10} got the first rank and it is indicated when a patient has shortness of breathing, the patient must go to treatment in the ICU. R_5 got the second rank by the technique VIKOR and R_{12} got the second place by TOPSIS technique. These are the next level symptoms of

Table 14 Finding the value of P_i and R_i

Risk factors	C_1	C_2	C_3	C_1	P_i	R_i
R_1	0.135	0.075	0.0409	0.0456	0.33145	0.11989
R_2	0.146	0.0825	0.0447	0.0456	0.0767	0.06143
R_3	0.135	0.0825	0.0447	0.0456	0.18113	0.10443
R_4	0.146	0.0825	0.0447	0.0456	0.0767	0.0767
R_5	0.135	0.06475	0.0409	0.0456	0.42603	0.20171
R_6	0.0986	0.06475	0.0409	0.03735	0.86642	0.45012
R_7	0.135	0.0825	0.0409	0.0456	0.22432	0.11989
R_8	0.135	0.06475	0.03825	0.03255	0.52432	0.15013
R_9	0.135001	0.075012	0.05145	0.0456	0.21156	0.10713
R_{10}	0.0986	0.0605	0.0447	0.03735	0.8715	0.45001
R_{11}	0.14601	0.06425	0.0447	0.0456	0.40398	0.20739
R_{12}	0.0986	0.0605	0.0409	0.0456	0.81989	0.45001

Table 15 Finding the value of P^*, R^*, P^- and R^-

Risk factors	P_i	R_i
R_1	0.33145	0.11989
R_2	0.0767	0.06143
R_3	0.18113	0.10443
R_4	0.0767	0.0767
R_5	0.42603	0.20171
R_6	0.86642	0.45000
R_7	0.22432	0.11989
R_8	0.52432	0.15000
R_9	0.21156	0.10713
R_{10}	0.8715	0.45000
R_{11}	0.40398	0.20739
R_{12}	0.81989	0.45000
P^*, R^*	0.0767	0.06143
P^-, R^-	0.8715	0.4500

COVID-19 and FLU, which are heavy sore throat and heavy body pain in the stage of the latent situation. R_2 got the last rank in both techniques, which is referred to as the risk factor cough is a normal symptom of more disease. So, the patient need not take any panic when they have a cough. Other risk factors such as fever, loss of taste or smell, and headache are getting median ranks. By this, we can estimate these risk factors are not dangerous or not ordinary symptoms. These are the early-level symptoms of any viral disease. Loss of speech is a very rare risk factor for most of the patients in the world. Diarrhoea is an unavoidable risk factor for babies who

Table 16 Finding the value of Q_i and find the ranking by descending order

Risk factors	Q_i	Rank
R_1	0.21914	8
R_2	0.0011	12
R_3	0.12145	10
R_4	0.01965	11
R_5	0.39826	2
R_6	0.2421	7
R_7	0.24391	6
R_8	0.39737	3
R_9	0.14419	9
R_{10}	1.002	1
R_{11}	0.39503	4
R_{12}	0.27638	5

Table 17 Comparing the ranking values of TOPSIS and VIKOR technique

Risk factors	VIKOR	TOPSIS
R_1	8	7
R_2	12	12
R_3	10	10
R_4	11	11
R_5	2	3
R_6	7	5
R_7	6	6
R_8	3	4
R_9	9	9
R_{10}	1	1
R_{11}	4	8
R_{12}	5	2

have the starting stage of immune system development. Hence, the risk factors of COVID-19 and FLU by the techniques of TOPSIS and VIKOR analysed in a simple way.

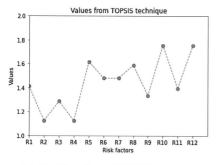
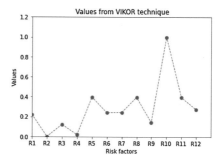

(a) Rankings from TOPSIS technique   (b) Rankings from VIKOR technique

Fig. 2 Rankings between TOPSIS and VIKOR

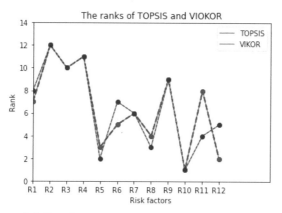

(a) Rankings from TOPSIS and VIKOR

6   Conclusion

In this work, a new hybrid TOPSIS and VIKOR technique is proposed to solve the MCDM problem by using fuzzy triangular numbers. In the proposed technique, the decision matrix form is created by using a linguistic form of fuzzy variables which can effectively represent the unavoidable uncertainty. This work proved the applicability of the MCDM methods in choosing the risk factors of COVID-19 and FLU on the opinion of doctors. Through the comparative study readers can analyse the risk factors of COVID-19 and FLU.

Acknowledgements The authors would like to thank Dr. Thiyagarajan and his team who are working in the general surgery department at Sathya Sai Medical College and Research Institute, Chennai. **Declarations** The authors declare no conflicts of interest in relation to this article.

References

1. Organization WH (2022) Considerations for integrating covid-19 vaccination into immunization programmes and primary health care for 2022 and beyond
2. Sotoudeh-Anvari A (2022) The applications of mcdm methods in covid-19 pandemic: A state of the art review. Appl Soft Comput 109238
3. Nguyen P-H, Tsai J-F, Dang T-T, Lin M-H, Pham H-A, Nguyen K-A (2021) A hybrid spherical fuzzy mcdm approach to prioritize governmental intervention strategies against the covid-19 pandemic: A case study from vietnam. Mathematics 9(20):2626
4. Alamoodi A, Zaidan B, Albahri O, Garfan S, Ahmaro IY, Mohammed R, Zaidan A, Ismail AR, Albahri A, Momani F, et al (2023) Systematic review of mcdm approach applied to the medical case studies of covid-19: trends, bibliographic analysis, challenges, motivations, recommendations, and future directions. Complex Intell Syst 1–27
5. Chowdhury NK, Kabir MA, Rahman MM, Islam SMS (2022) Machine learning for detecting covid-19 from cough sounds: An ensemble-based mcdm method. Comput Biol Med 145:105405
6. Wise J (2021) Vaccinating against covid and flu at same time is safe, study shows. British Med J Publ Group
7. Nicola J (2020) How covid-19 is changing the cold and flu season. Nature 588(7838):388–390
8. Kiseleva I, Ksenafontov A (2021) Covid-19 shuts doors to flu but keeps them open to rhinoviruses. Biology 10(8):733
9. Jaklevic MC (2020) Flu vaccination urged during covid-19 pandemic. Jama 324(10):926–927
10. Gasmi A, Peana M, Pivina L, Srinath S, Benahmed AG, Semenova Y, Menzel A, Dadar M, Bjørklund G (2021) Interrelations between covid-19 and other disorders. Clin Immunol 224:108651
11. Onakpojeruo EP, Uzun B, Ozsahin I, Ozsahin DU (2022) Evaluation of the treatment alternatives for spinal cord tumors using analytical evaluation models
12. Ecer F (2022) An extended mairca method using intuitionistic fuzzy sets for coronavirus vaccine selection in the age of covid-19. Neural Comput Appl 34(7):5603–5623
13. Yildirim FS, Sayan M, Sanlidag T, Uzun B, Ozsahin DU, Ozsahin I (2021) Comparative evaluation of the treatment of covid-19 with multicriteria decision-making techniques. J Health Eng 2021
14. Alsalem M, Alamoodi A, Albahri O, Dawood K, Mohammed R, Alnoor A, Zaidan A, Albahri A, Zaidan B, Jumaah F et al (2022) Multi-criteria decision-making for coronavirus disease 2019 applications: a theoretical analysis review. Artif Intell Rev 55(6):4979–5062
15. Ali SA, Ahmad A (2019) Spatial susceptibility analysis of vector-borne diseases in kmc using geospatial technique and mcdm approach. Model Earth Syst Environ 5:1135–1159
16. Mohammed K, Jaafar J, Zaidan A, Albahri OS, Zaidan B, Abdulkareem KH, Jasim AN, Shareef AH, Baqer M, Albahri AS et al (2020) A uniform intelligent prioritisation for solving diverse and big data generated from multiple chronic diseases patients based on hybrid decision-making and voting method. IEEE Access 8:91521–91530
17. Yas QM et al (2021) Evaluation multi diabetes mellitus symptoms by integrated fuzzy-based mcdm approach. Turkish J Comput Math Edu (TURCOMAT) 12(13):4069–4082
18. Hwang CL, Yoon K, Hwang CL, Yoon K (1981) Methods for multiple attribute decision making. Multiple attribute decision making: methods and applications a state-of-the-art survey 58–191
19. Naeem K, Riaz M, Karaaslan F (2021) A mathematical approach to medical diagnosis via pythagorean fuzzy soft topsis, vikor and generalized aggregation operators. Complex Intell Syst 7:2783–2795
20. Akram M, Arshad M (2020) Bipolar fuzzy topsis and bipolar fuzzy electre-i methods to diagnosis. Comput Appl Math 39:1–21
21. Mohammed MA, Abdulkareem KH, Al-Waisy AS, Mostafa SA, Al-Fahdawi S, Dinar AM, Alhakami W, Abdullah B, Al-Mhiqani MN, Alhakami H et al (2020) Benchmarking methodology for selection of optimal covid-19 diagnostic model based on entropy and topsis methods. IEEE Access 8:99115–99131

22. Zadeh LA (1965) Fuzzy sets. Inf. Control 8(3):338–353
23. Dubois D, Prade H (1987) The mean value of a fuzzy number. Fuzzy Sets Syst 24(3):279–300
24. Pedrycz W (1994) Why triangular membership functions? Fuzzy Sets Syst 64(1):21–30
25. Zadeh LA (1996) A computational theory of dispositions. In: Fuzzy sets, fuzzy logic, and fuzzy systems: selected papers by Lotfi A Zadeh. World Scientific, pp 713–737
26. Molla MU, Giri BC, Biswas P (2021) Extended promethee method with pythagorean fuzzy sets for medical diagnosis problems. Soft Comput 25:4503–4512
27. Liao Z, Liao H, Lev B (2022) Compromise solutions for stochastic multicriteria acceptability analysis with uncertain preferences and nonmonotonic criteria. Int Trans Oper Res 29(6):3737–3757
28. Chinnasami Sivaji MR, Kurinjimalar Ramu SS (2021) A review on weight process method and its classification. Data Anal Artif Intell 1(1):1–8
29. El Sayed M, El Safty M, El-Bably M (2021) Topological approach for decision-making of covid-19 infection via a nano-topology model. AIMS Math 6(7):7872–7894
30. Sun P, Qie S, Liu Z, Ren J, Li K, Xi J (2020) Clinical characteristics of hospitalized patients with sars-cov-2 infection: a single arm meta-analysis. J Med Virol 92(6):612–617
31. Caronna E, Pozo-Rosich P (2021) Headache as a symptom of covid-19: narrative review of 1-year research. Curr Pain Head Rep 25(11):73
32. Ghimire S, Sharma S, Patel A, Budhathoki R, Chakinala R, Khan H, Lincoln M, Georgeston M (2021) Diarrhea is associated with increased severity of disease in covid-19: systemic review and metaanalysis. SN Comprehen Clin Med 3:28–35
33. Suresh S, Modi R, Sharma A, Arisutha S, Sillanpää M (2021) Pre-covid-19 pandemic: effects on air quality in the three cities of india using fuzzy mcdm model. J Environ Health Sci Eng 1–11
34. Ahmad S, Masood S, Khan NZ, Badruddin IA, Ahmadian A, Khan ZA, Khan AH et al (2023) Analysing the impact of covid-19 pandemic on the psychological health of people using fuzzy mcdm methods. Oper Res Perspect 10:100263
35. Homans NC, Vroegop JL (2022) The impact of face masks on the communication of adults with hearing loss during covid-19 in a clinical setting. Int J Audiol 61(5):365–370

Robust Numerical Technique for a Class of Singularly Perturbed Nonlinear System of n-Differential Equations With Robin Boundary Conditions

R. Ishwariya⊙

Abstract In this article, a class of nonlinear singularly perturbed system of n-differential equations with Robin boundary conditions of unperturbed type is considered. The numerical method considered in this work consists of the classical finite difference operator over a piecewise uniform Shishkin mesh together with a continuation algorithm. The method suggested in this paper is proved to be essentially first-order convergent uniformly with respect to all perturbation parameters. Numerical experiments are carried out for Robin boundary conditions with and without perturbation parameters.

Keywords Singular perturbation problems · Nonlinear differential equations · Robin boundary conditions · Shishkin mesh · Finite difference scheme · Parameter-uniform convergence

1 Introduction

A differential equation(DE) whose higher and/or lower order derivatives got multiplied by a small perturbation parameter say $\varepsilon \ll 1$ is known as singularly perturbed differential equation (SPDE). Due to the presence of this perturbation parameter, the solution contains a region(s) of rapid variation is\are called intial\boundary layers. Classical numerical methods suffer major defects in these regions.

In this paper, a class of nonlinear singularly perturbed systems of differential equations of reaction-diffusion type for different perturbation parameters with Robin boundary conditions is considered. Numerical solutions are obtained by considering the finite difference scheme on Shishkin mesh and the numerical method suggested is proved to be essentially first-order convergent independent of the perturbation parameters. Due to the absence of the perturbation parameters in the boundary

R. Ishwariya (✉)
Department of Science and Humanities, Amrita School of Engineering, Amrita Vishwa
Vidyapeetham, Chennai 601103, Tamil Nadu, India
e-mail: r_ishwariya@ch.amrita.edu

© The Author(s), under exclusive license to Springer Nature Singapore Pte Ltd. 2024
D. Giri et al. (eds.), *Proceedings of the Tenth International Conference on Mathematics and Computing*, Lecture Notes in Networks and Systems 963,
https://doi.org/10.1007/978-981-97-2069-9_13

conditions, the solution $\vec{y}$ exhibit less severe boundary layers compared to problems with perturbed boundary conditions. Also, for the problem with unperturbed boundary conditions, the first derivative $\vec{y}'$ of the solution $\vec{y}$ exhibit strong boundary layers which are also reflected in the bounds of the solutions and their derivatives. Numerical comparison is done for problems with and without perturbation parameters in the boundary conditions.

In [2], the behavior of a linear singularly perturbed convection-diffusion DE with Robin boundary conditions is studied and a numerical method consisting of an upwind finite difference scheme on Shishkin meshes is developed and it is proved to be uniformly convergent independent of the diffusion coefficient. In [3], a nonlinear second-order singularly perturbed convection-diffusion type DE satisfying the Robin boundary conditions is considered and two novel algorithms are developed for determining the solution. They introduced a boundary shape function for the transformation of the BVP into an IVP for the new variable by two different types of algorithms.

In [5], a parameter uniform numerical method with an appropriate layer-adapted piecewise uniform mesh is constructed for a semilinear system of singularly perturbed DEs of reaction-diffusion type. In [6], a singularly perturbed nonlinear BVPs of reaction diffusion type is studied and an exponentially fitted difference scheme on a uniform mesh is developed. In [7], a class of two parameters singularly perturbed nonlinear reaction-diffusion type DE with initial and boundary value conditions is considered. The asymptotic behaviors of the solution are discussed by the construction of the asymptotic expansion of the solution under suitable conditions.

In [8], Ishwariya et.al. studied the biochemical reaction namely Michaelis-Menten kinetics which is mathematically modeled into a system of nonlinear singularly perturbed first-order IVPs. The numerical method suggested is proved to be parameter uniform almost first-order convergent in the maximum norm. In [9], Ishwariya et.al. considered a class of parabolic linear systems of second-order singularly perturbed differential equations of reaction-diffusion type with initial and Robin boundary conditions. A classical finite difference scheme constructed on a piecewise uniform mesh is suggested and the numerical method is proved to be first-order convergent and essentially first-order convergent in the time and space variable respectively in the maximum norm independent of the perturbation parameters. Also, due to the absence of the perturbation parameters in the boundary conditions, the solution components $\vec{u}$ exhibit smooth layers, whereas the first derivatives $\dfrac{\partial \vec{u}}{\partial x}$ of the solution exhibit parabolic boundary layers.

In [10], singularly perturbed boundary value problems for semilinear reaction-diffusion equations were investigated. A uniform mesh with Numerov's type scheme is considered. The method suggested is proved to be uniformly convergent, with respect to the discrete maximum norm, independently of the parameter of ε. In [11], the multiple solutions for which boundary or interior layers exhibit are discussed for semilinear BVPs of reaction-diffusion type. In [12], O'Malley a nonlinear system of partially singularly perturbed IVPs is studied. In [14], the existence, uniqueness, and asymptotic estimates of solutions are discussed for a second-order nonlinear singularly perturbed convection-diffusion type ODE with Robin boundary conditions.

It is to be noted that in the present work, no artificial conditions on the perturbation parameters are imposed.

A class of nonlinear systems of singularly perturbed BVP with Robin boundary conditions is considered as follows:

$$-\mathcal{E}\vec{y}''(x) + \vec{g}(x, \vec{y}) = 0 \text{ on } \mathcal{D} = (0, 1) \tag{1}$$

with

$$\vec{y}(0) - \vec{y}'(0) = \vec{\phi} \text{ and } \vec{y}(1) + \vec{y}'(1) = \vec{\psi}, \tag{2}$$

where the vectors $\vec{\phi}$ and $\vec{\psi}$ are constants given, $\mathcal{E}$ is a $n \times n$ matrix, $\mathcal{E} = \text{diag}(\vec{\varepsilon})$, $\vec{\varepsilon} = (\vec{\varepsilon}_1, \vec{\varepsilon}_2, \ldots, \vec{\varepsilon}_n)$ For all $(x, \vec{y}) \in \bar{\mathcal{D}} \times \mathbb{R}^n$ and $\vec{g}(x, \vec{y}) \in C^4(\bar{\mathcal{D}} \times \mathbb{R}^n)$ the following condition is assumed that the nonlinear terms satisfy

$$\frac{\partial g_k}{\partial y_k} \geq \beta > 0, \quad \frac{\partial g_k}{\partial y_j} \leq 0, \quad k, j = 1, \ldots, n, \quad k \neq j, \tag{3}$$

$$\min_{1 \leq i \leq n} \left(\sum_{j=1}^{n} \frac{\partial g_i(x, y)}{\partial y_j} \right) \geq \alpha > 0, \text{ for some constant } \alpha \tag{4}$$

where $\bar{\mathcal{D}} = [0, 1]$. The implicit function theorem along with the assumption (4) ensures that $\vec{y} \in C^4(\bar{\mathcal{D}})$. The operator form of the problem (1)–(2) can be written as follows:

$$T\vec{y}(x) := -\mathcal{E}\vec{y}''(x) + \vec{g}(x, \vec{y}) = 0 \text{ on } \mathcal{D} \tag{5}$$

with

$$B_0\vec{y}(0) := \vec{\phi} \text{ and } B_1\vec{y}(1) := \vec{\psi} \tag{6}$$

where $B_0 = I - \dfrac{d}{dx}$ and $B_1 = I + \dfrac{d}{dx}$.

Each component y_l, $l = 1, \ldots, n$ of the solution $\vec{y}$ are expected to have weak twin layers of width $O(\sqrt{\varepsilon_n})$ at $x = 0$ and $x = 1$. In addition, the components y_l, $l = 1, \ldots, n - 1$ exhibits weak twin layers of width $O(\sqrt{\varepsilon_{n-1}})$, the components y_l, $l = 1, \ldots, n - 2$ form weak twin boundary layers of width $O(\sqrt{\varepsilon_{n-2}})$ and so on.

Note that, the positive constant C is not depending on x, ε_i and N, the discretization parameter throughout the paper.

2 Analytical Results

The reduced problem obtained by putting $\vec{\varepsilon} = 0$ in (1)–(2) is given by

$$\vec{g}(x, \vec{r}) = 0 \text{ on } \mathcal{D}. \tag{7}$$

The existence and uniqueness of the solution for (7) is ensured by the condition (4) and the implicit function theorem. Further, it is to be noted, that the bounds of the solutions r_i of (7) and their derivatives are independent of ε_i. Hence,

$$|r_i^{(m)}(x)| \le C \quad \text{for } i = 1, \ldots, n, \; m = 0, 1, 2, 3 \text{ and } x \in \bar{\mathcal{D}}. \tag{8}$$

The solution $\vec{y}(x)$ of (1)–(2) is decomposed into a smooth component $\vec{p}(x)$ and a singular component $\vec{q}(x)$ as follows:

$$\vec{y}(x) = \vec{p}(x) + \vec{q}(x)$$

where

$$T\vec{p}(x) := -\mathcal{E}p''(x) + \vec{g}(x, \vec{p}) = 0 \text{ on } \mathcal{D} \tag{9}$$

with

$$B_0 \vec{p}(0) = B_0 \vec{r}(0) \text{ and } B_1 \vec{p}(1) = B_1 \vec{r}(1) \tag{10}$$

and

$$T\vec{q}(x) := -\mathcal{E}q''(x) + \vec{g}(x, \vec{p} + \vec{q}) - \vec{g}(x, \vec{p}) = 0 \text{ on } \mathcal{D} \tag{11}$$

with

$$B_0 \vec{q}(0) = B_0(\vec{y} - \vec{p})(0) \text{ and } B_1 \vec{q}(1) = B_1(\vec{y} - \vec{p})(1). \tag{12}$$

Theorem 1 *For all $i = 1, \ldots, n$ and $x \in \bar{\mathcal{D}}$,*

$$|p_i^{(k)}(x)| \le C, \text{ for } k = 0, 1, 2, 3 \text{ and } |p_i^{(4)}(x)| \le C(1 + \varepsilon_i^{-1/2}).$$

Proof The smooth component $\vec{p}(x)$ is further decomposed as follows:

$$\vec{p}(x) = \sum_{i=1}^{n} \vec{v}^{[i]}(x)$$

where $\vec{v}^{[n]}(x)$ is the solution of

$$-\mathcal{E}_n \frac{d^2 \vec{v}^{[n]}}{dx^2} + f(x, v_1^{[j]}, v_2^{[j]}, \ldots, v_n^{[j]}) = 0, \quad x \in \mathcal{D}, \tag{13}$$

$$B_0 \vec{v}^{[n]}(0) = B_0 p(0) \text{ and } B_1 \vec{v}^{[n]}(1) = B_1 p(1) \tag{14}$$

and $\vec{v}^{[i]}(x)$, $i = 1, 2, \ldots, n$ are the solutions of

$$- \mathcal{E}_i \frac{d^2 \vec{v}^{[i]}}{dx^2} + \vec{f}(x, \sum_{j=i}^{n} v_1^{[j]}, \ldots, \sum_{j=i}^{n} v_n^{[j]}) - \vec{f}(x, \sum_{j=i+1}^{n} v_1^{[j]}, \ldots, \sum_{j=i+1}^{n} v_n^{[j]}) = 0, \quad x \in \mathcal{D},$$

(15)

$$B_0 \vec{v}^{[i]}(0) = 0 \text{ and } B_1 \vec{v}^{[i]}(1) = 0. \tag{16}$$

where $\mathcal{E}_i$ is the diagonal matrix $diag(0, 0, \ldots, \varepsilon_i, \varepsilon_{i+1}, \ldots, \varepsilon_n)$, $i = 1, 2, \ldots, n$.
Let $x \in \mathcal{D}$. Using (7) and (13), we get

$$- \mathcal{E}_n \frac{d^2 \vec{v}^{[n]}(x)}{dx^2} + A^{[n]}(x)(\vec{v}^{[n]} - r)(x) = \mathcal{E}_n \frac{d^2 \vec{r}(x)}{dx^2} \tag{17}$$

where $A^{[n]}(x) = (a_{ij}^{[n]}(x))$ with $a_{ij}^{[n]}(x) = \dfrac{\partial f_i}{\partial y_j}(x, \vec{\chi}_i^{[n]}(x))$ is the intermediate value.

Consider the linear operator,

$$T_1 z(x) = -\varepsilon_n z''(x) + a_1(x)z(x) = \varepsilon_n r_n''(x) \tag{18}$$

where $z = v_n^{[n]} - r_n$ and $a_1(x)$ is a rational function in $a_{ij}^{[n]}(x)$ derived in expressing $v_i^{[n]} - r^i$ in terms of $v_n^{[n]} - r_n$.

From (10) and (14), we drive

$$B_0 z(0) = 0 \text{ and } B_1 z(1) = 0. \tag{19}$$

The operator T_1 together with (19) satisfies the maximum principle in [13].
Thus,

$$|z(x)| \leq C\varepsilon_n. \tag{20}$$

On differentiating (18) once, we get

$$T_1 z'(x) = -\varepsilon_n z'''(x) + a^*(x)z'(x) = \varepsilon_n r_n'''(x) - a^{*'}(x)z(x). \tag{21}$$

Rearranging (19), we get

$$z'(0) = z(0) \text{ and } z'(1) = -z(1). \tag{22}$$

Denoting z' by h in (21) and (22), we get

$$T_1 h(x) = -\varepsilon_n h''(x) + a^*(x)h(x) = \varepsilon_n r_n'''(x) - a^{*'}(x)z(x), \tag{23}$$

$$h(0) = z(0) \text{ and } h(1) = -z(1). \tag{24}$$

Problem (23)–(24) satisfies the maximum principle in [13]. Thus,

$$|h(x)| \leq C\varepsilon_n. \tag{25}$$

Rearranging (23) and using (25), we get

$$|h''(x)| \leq C. \tag{26}$$

From the mean-value theorem, we get

$$|h'(x)| \leq C\varepsilon_n^{1/2}. \tag{27}$$

Differentiating (23) once and rearranging, we get

$$|h'''(x)| \leq C(1 + \varepsilon_n^{-1/2}). \tag{28}$$

Thus from (25)–(28), we get

$$|\frac{d^k v_n^{[n]}(x)}{dx^k}| \leq C, \text{ for } k = 0, 1, 2, 3 \text{ and } |\frac{d^4 v_n^{[n]}(x)}{dx^4}| \leq C(1 + \varepsilon_n^{-1/2}).$$

Also from the remaining $n - 1$ equations of the system (13), we get

$$|\frac{d^k v_l^{[n]}(x)}{dx^k}| \leq C, \text{ for } k = 0, 1, 2, 3$$

and

$$|\frac{d^4 v_l^{[n]}(x)}{dx^4}| \leq C(1 + \varepsilon_n^{-1/2}) \leq C(1 + \varepsilon_l^{-1/2}), \quad l = 1, \dots, n - 1.$$

Using similar arguments in the system (15), we derive

$$|\frac{d^k v_l^{[n]}(x)}{dx^k}| \leq C, \text{ for } k = 0, 1, 2, 3 \text{ and } |\frac{d^4 v_l^{[n]}(x)}{dx^4}| \leq C(1 + \varepsilon_l^{-1/2}) \ l = 1, \dots, n.$$

Thus, the bounds of $\vec{v}^{[i]}$ and their derivatives give the required bounds for $\vec{p}$ and its derivatives.

The layer functions $\mathcal{B}_i^L$, $\mathcal{B}_i^R$, $\mathcal{B}_i$, $i = 1, \dots, n$, associated with the solution $\vec{u}$, are defined on $\bar{D}$ by

$$\mathcal{B}_i^L(x) = e^{-x\sqrt{\alpha/\varepsilon_i}}, \ \mathcal{B}_i^R(x) = \mathcal{B}_i^L(1 - x), \ \mathcal{B}_i(x) = \mathcal{B}_i^L(x) + \mathcal{B}_i^R(x).$$

Theorem 2 *For all* $i = 1, \ldots, n$ *and* $x \in \bar{\mathcal{D}}$,

$$|q_i^{(k)}(x)| \leq C \mathcal{B}_n(x), \quad k = 0, 1,$$

$$|q_i^{(k)}(x)| \leq C \sum_{m=i}^{n} \varepsilon_m^{-\frac{(k-1)}{2}} \mathcal{B}_m(x), \quad k = 2, 3,$$

$$|\varepsilon_i q_i^{(4)}(x)| \leq C \sum_{m=1}^{n} \varepsilon_m^{-\frac{1}{2}} \mathcal{B}_m(x).$$

Proof From (11), we get

$$- E\vec{q}''(x) + B(x)\vec{q}(x) = \vec{0} \tag{29}$$

where $B(x) = (b_{ij}(x))$ with $b_{ij}(x) = \dfrac{\partial f_i}{\partial y_j}(x, \vec{\lambda}_i(x))$ is the intermediate value.
The proof is analogs to the proof of Lemma 3.6 in [9]. Thus the bounds for the singular component $\vec{q}$ and its derivatives hold.

3   Shishkin Mesh

In this section, a special type of piecewise uniform mesh known as Shishkin mesh is constructed on $\bar{\mathcal{D}}$ with N number of mesh-intervals is given as follows. Let $\mathcal{D}^N = \{x_j\}_{j=1}^{N}$ and $\bar{\mathcal{D}}^N = \{x_j\}_{j=0}^{N}$. The domain $\bar{\mathcal{D}}$ is divided into $2n + 1$ intervals as follows:

$$[0, \tau] \cup (\tau_{n-1}, \tau_n] \cup \cdots \cup (\tau_n, 1 - \tau_n] \cup (1 - \tau_n, 1 - \tau_{n-1}] \cup \cdots \cup (1 - \tau, 1].$$

The n parameters τ_r, $r = n - 1, \ldots, 1$ are defined by

$$\tau_n = \min \left\{ \frac{1}{4}, 2 \frac{\sqrt{\varepsilon_n}}{\sqrt{\alpha}} \ln N \right\},$$

$$\tau_r = \min \left\{ \frac{r \tau_{r+1}}{r + 1}, 2 \frac{\sqrt{\varepsilon_r}}{\sqrt{\alpha}} \ln N \right\}.$$

On the sub-interval $(\tau_{n-1}, \tau_n]$, $\dfrac{N}{2}$ mesh-intervals is placed to form a uniform mesh and on each of the mesh-intervals $(\tau_r, \tau_{r+1}]$ and $(1 - \tau_{r+1}, 1 - \tau_r]$, $r = 0, 1, \ldots, n - 1$, $\dfrac{N}{4n}$ mesh-intervals is considered to form a uniform mesh. For practical convenience, we take $N = 4nk$, $k \geq 3$, where $'n'$ denotes the number of distinct singular perturbation parameters involved in the given problem.

4 Discrete Problem

The discrete problem corresponding to the continuous problem (1)–(2) is defined as follows:

$$T^N \vec{y}(x_j) := -\mathcal{E}\delta^2 \vec{y}(x_j) + \vec{g}(x_j, Y(x_j)) = 0, \quad \text{for } x_j \in \mathcal{D}^N \qquad (30)$$

with

$$B_0^N \vec{y}(0) = B_0 \vec{y}(0) \text{ and } B_1^N \vec{y}(1) = B_1 \vec{y}(1) \qquad (31)$$

where $B_0^N = I - D^+$, $B_1^N = I + D^-$.

Here, $\quad D^+ Y(x_j) = \dfrac{Y(x_{j+1}) - Y(x_j)}{h_j}, \quad D^- Y(x_j) = \dfrac{Y(x_j) - Y(x_{j-1})}{h_j}$ and

$\delta^2 Y(x_j) = \dfrac{Y(x_{j+1}) - Y(x_{j-1})}{2h_j}, \; h_j = x_j - x_{j-1}.$

5 Error Analysis

Theorem 3 *For given any two discrete mesh functions $\vec{u}$ and $\vec{W}$ with $B_0^N \vec{u}(0) = B_0^N \vec{W}(0)$ and $B_1^N \vec{u}(1) = B_1^N \vec{W}(1)$,*

$$|\vec{u} - \vec{W}| \leq C \, |T^N(\vec{u} - \vec{W})|.$$

Proof

$$(T^N(\vec{u} - \vec{W}))(x_j)$$

$$= -E\,\delta^2(\vec{U} - \vec{W})(x_j) + \vec{g}(x_j, \vec{U}(x_j)) - f(x_j, \vec{W}(x_j))$$

$$= -E\,\delta^2(\vec{U} - \vec{W})(x_j) + \frac{\partial \vec{g}}{\partial y}(x_j, \vec{M}(x_j))(\vec{U} - \vec{W})(x_j) \qquad (32)$$

$$= T^{N^*}(\vec{U} - \vec{W})(x_j)$$

where $\dfrac{\partial \vec{g}}{\partial y}(x_j, \vec{M}(x_j))$ is the intermediate value and T^{N^*} denotes the Frechet derivative of $\acute{T}^N$.

Note that, T^{N^*} is a linear operator and hence satisfies the following discrete maximum principle and discrete stability result corresponding to the linear problem,

$$T^{N^*} := -\mathcal{E}\delta^2 \vec{y}(x_j) + A(x_j)\vec{y}(x_j) = \vec{g}(x_j), \quad \text{for } x_j \in \mathcal{D}^N \qquad (33)$$

with

$$B_0^* \vec{Y}(0) := \vec{Y}(0) - D^+ \vec{Y}(0) = \vec{\phi}, \ B_1^* \vec{Y}(1) := \vec{Y}(1) + D^- \vec{Y}(1) = \vec{\psi}. \quad (34)$$

Lemma 1 *Let* $\vec{\Psi}$ *be any vector-valued mesh function, such that* $B_0 \vec{\Psi}(0) \geq \vec{0}$, $B_1 \vec{\Psi}(1) \geq \vec{0}$. *Then* $T^{N*} \vec{\Psi} \geq \vec{0}$ *on* $\mathcal{D}^N$ *implies that* $\vec{\Psi}(1) \geq \vec{0}$ *on* $\bar{\mathcal{D}}^N$.

Lemma 2 *For any vector-valued mesh function* $\vec{\Psi}$ *on* $\mathcal{D}^N$, *then for each* $i = 1, \ldots, n$,

$$|\Psi_i(x_j)| \leq \max ||B_0^N \vec{\Psi}(0)||, \ ||B_1^N \vec{\Psi}(0)||, \ \frac{1}{\alpha}||T^{N*} \vec{\Psi}||, \ 0 \leq j \leq N.$$

Hence, on $\mathcal{D}^N$

$$|\vec{U} - \vec{W}| \ \leq C \, |T^{N*}(\vec{U} - \vec{W})| \ = C \, |T^N(\vec{U} - \vec{W})|. \quad (35)$$

Theorem 4 *Let* $\vec{y}$ *be the solution of* (1)–(2) *and* $\vec{y}$ *be the solution of* (30)–(31). *Then for* $x_j \in \bar{\mathcal{D}}^N$,

$$|(\vec{y} - \vec{y})(x_j)| \ \leq \ C \, N^{-1} \ln N. \quad (36)$$

Proof Let $x_j \in \mathcal{D}^N$. Since $B_0^N \vec{y}(0) = B_0 \vec{y}(0)$ and $B_1^N \vec{y}(1) = B_1 \vec{y}(1)$ from (35),

$$|Y - y| \ \leq \ C \, |T^N(\vec{y} - \vec{y})|.$$

Using (30), we get

$$|T^N \vec{y}(x_j)| \ = \ |(T^N \vec{y} - T^N \vec{y})(x_j)|.$$

Consider,

$$|(T^N \vec{y} - T^N \vec{y})(x_j)| = |T^N \vec{y}(x_j)|$$

$$= |(T^N \vec{y} - T \vec{y})(x_j)|$$

$$= E|(\delta^2 - D^2)\vec{y}(x_j)|$$

$$\leq E(|(\delta^2 - D^2)\vec{p}(x_j)| + |(\delta^2 - D^2)\vec{q}(x_j)|)$$

where $D^2 = \frac{d^2}{dx^2}$. From the results of local truncation error in [13], the following result holds.

$$|(T^N(\vec{y} - \vec{y}))(x_j)| \leq C N^{-1} \ln N.$$

Thus,

$$|(\vec{y} - \vec{y})(x_j)| \leq C N^{-1} \ln N.$$

6 Numerical Illustrations

In this section, two numerical examples are discussed. In the first example, unperturbed Robin boundary conditions are considered, whereas in the second example, the perturbation parameters ε_i also occur in the Robin boundary conditions. To solve the vector problems in the examples, the continuation algorithm which is a variant of the one found in [4] is considered. The rate of convergence, the maximum pointwise error and the error constant, which are all parameter uniform are denoted by the notations p^N, D^N and C_p^N respectively as given in [4].

Example 1 Consider the following nonlinear BVP with unperturbed Robin boundary conditions

$$-\varepsilon_1 y_1''(x) + y_1^3(x) + 2\,y_1(x) - 0.1\,y_2(x) = 0,$$
$$-\varepsilon_2 y_2''(x) + y_2^3(x) + 2\,y_2(x) - 0.1\,y_1(x) = 0, \quad x \in (0, 1)$$

with $y_i(0) - y_i'(0) = 1.0$ and $y_i(1) + y_i'(1) = 1.0$, $i = 1, 2$.

Example 2 Consider the following nonlinear BVP with Robin boundary conditions which include $\sqrt{\varepsilon_i}$

$$-\varepsilon_1 y_1''(x) + y_1^3(x) + 2\,y_1(x) - 0.1\,y_2(x) = 0,$$
$$-\varepsilon_2 y_2''(x) + y_2^3(x) + 2\,y_2(x) - 0.1\,y_1(x) = 0, \quad x \in (0, 1)$$

with $y_i(0) - \sqrt{\varepsilon_i}\,y_i'(0) = 1.0$ and $y_i(1) + \sqrt{\varepsilon_i}\,y_i'(1) = 1.0$, $i = 1, 2$.

The maximum pointwise errors and the rate of convergence for the above BVPs are presented in Tables 1 and 2 and graphs of the numerical solutions of $y_1(x)$ and $y_2(x)$ for $N = 1024$ and $\varepsilon = 2^{-9}$ and 2^{-8} are portrayed in Figs. 1 and 3.

Table 1 Values of D^N, p^N and C_p^N for $\varepsilon_1 = \dfrac{\eta}{16}$, $\varepsilon_2 = \dfrac{\eta}{8}$ and $\alpha = 1.899$

η	Number of mesh points N			
	64	128	256	512
2^{-2}	0.47217E-02	0.23503E-02	0.11731E-02	0.58614E-03
2^{-4}	0.47748E-02	0.23872E-02	0.11945E-02	0.59771E-03
2^{-6}	0.47893E-02	0.24369E-02	0.12319E-02	0.61955E-03
2^{-8}	0.48750E-02	0.24876E-02	0.12611E-02	0.63539E-03
2^{-10}	0.49514E-02	0.25183E-02	0.12800E-02	0.64644E-03
D^N	0.49514E-02	0.25183E-02	0.12800E-02	0.64644E-03
p^N	0.97537E+00	0.97629E+00	0.98560E+00	
C_p^N	0.58210E+00	0.58210E+00	0.58173E+00	0.57762E+00

Table 2 Values of D^N, p^N and C_p^N for $\varepsilon_1 = \dfrac{\eta}{16}$, $\varepsilon_2 = \dfrac{\eta}{8}$ and $\alpha = 1.899$

η	Number of mesh points N			
	64	128	256	512
2^{-2}	0.12857E-01	0.66366E-02	0.33718E-02	0.16995E-02
2^{-4}	0.17684E-01	0.92482E-02	0.47298E-02	0.23918E-02
2^{-6}	0.24085E-01	0.12828E-01	0.66225E-02	0.33648E-02
2^{-8}	0.32344E-01	0.17662E-01	0.92387E-02	0.47254E-02
2^{-10}	0.42652E-01	0.24067E-01	0.12821E-01	0.66193E-02
D^N	0.42652E-01	0.24067E-01	0.12821E-01	0.66193E-02
p^N	0.82555E+00	0.90857E+00	0.95374E+00	
C_p^N	0.30325E+01	0.30325E+01	0.28629E+01	0.26195E+01

Fig. 1 Numerical approximations of $y_1(x)$ and $y_2(x)$ in Example 1 for $\varepsilon_1 = 2^{-9}$, $\varepsilon_2 = 2^{-8}$

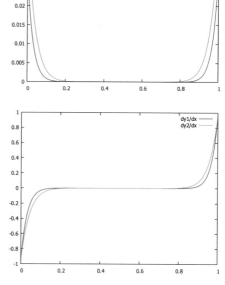

Fig. 2 Numerical approximations of dy_1/dx and dy_2/dx in Example 1 for $\varepsilon_1 = 2^{-9}$, $\varepsilon_2 = 2^{-8}$

7   Conclusion

From Figs. 1 and 2, it is observed that the solutions y_1 and y_2 exhibit weak boundary layers due to the absence of singular perturbation parameters in the boundary conditions and their derivatives y_1' and y_2' have strong boundary layers. Whereas, from Fig. 3, we note that the solutions y_1 and y_2 exhibit strong boundary layers since the

Fig. 3 Numerical
approximations of $y_1(x)$ and
$y_2(x)$ in Example 2 for
$\varepsilon_1 = 2^{-9}$, $\varepsilon_2 = 2^{-8}$

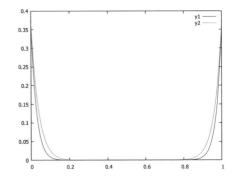

Fig. 4 Comparison of the
numerical approximations in
Examples 1 and 2 for
$\varepsilon_1 = 2^{-9}$, $\varepsilon_2 = 2^{-8}$

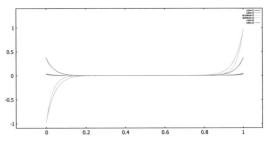

boundary conditions are perturbed. A comparison of the severity of the boundary
layers of the solutions and their derivatives due to the presence and the absence of
the perturbation parameters ε_i are given the Fig. 4.

From the tables, we find that as the η value decreases to zero and the number of
mesh points increases, the maximum pointwise errors decrease diagonally and the
numerical method proposed is essentially first-order parameter uniform convergent.
Further, it is observed from the tables the parameter uniform error constant decreases
monotonically.

The computational technique suggested in this paper is both robust and layer-
resolving. Further, the present computational technique is also applicable to problems
of the same class with Neumann boundary conditions and also with a much wider
class of singularities. In the future, the same class of problems with partially perturbed
systems can be discussed and hybrid numerical methods can also be developed for
higher-order convergence.

References

1. Miller JJH, O'Riordan E, Shishkin GI (1996) Fitted numerical methods for singular perturbation
   problems. World Scientific Publishing Co., Singapore, New Jersey, London, Hong Kong
2. Ansari Alan A, Hegarty F (2003) Numerical solution of a convection diffusion problem with
   robin boundary conditions. J Comput Appl Math 156(1):221–238

3. Chein-Shan L, Jiang-Ren C (2020) Boundary shape functions methods for solving the nonlinear singularly perturbed problems with Robin boundary conditions. Int J Nonlinear Sci Numer Simul 21(7–8):797–806
4. Farrell PA, Hegarty AF, Miller JJH, Riordan EO, Shishkin GI (2000) Robust computational techniques for boundary layers. Chapman and hall/CRC, Boca Raton, Florida, USA
5. Gracia JL, Lisbona FJ, O'Riordan E (2010) A coupled system of singularly perturbed parabolic reaction-diffusion equations. Adv Comput Math 32(1):43–61
6. Hakki D, Baransel G (2019) Numerical solutions for singularly perturbed nonlinear reaction diffusion boundary value problems. IOSR J Math 15(1):35–49
7. Huaijun C, Jiaqi M (2010) Singularly perturbed nonlinear reaction diffusion problem with two parameters. Acta Phys Sinica -Chinese Edn 59(7)
8. Ishwariya R, Princy Merlin J, Miller JJH, Valarmathi S (2016) A parameter uniform almost first order convergent numerical method for a nonlinear system of singularly perturbed differential equations. Biomath 5:1608111
9. Ishwariya R, Miller JJH, Valarmathi S (2019) A parameter uniform essentially first-order convergent numerical method for a parabolic system of singularly perturbed differential equations of reaction-diffusion type with initial and Robin boundary conditions. Int J Biomath 12(1):1950001
10. Yamac K, Erdogan F (2020) A numerical scheme for semilinear singularly perturbed reaction-diffusion problems. Appl Math Nonlinear Sci 5(1):405–412
11. Stynes M, Kopteva N (2006) Numerical analysis of singularly perturbed nonlinear reaction-diffusion problems with multiple solutions. Comput Math Appl 51(5):857–864
12. O'Malley RE (1988) On nonlinear singularly perturbed initial value problems. SIAM 30(2):193–212
13. Valarmathi S (2003) Numerical methods for singularly perturbed boundary value problems for third order ordinary differential equations. Doctoral Thesis, Bharathidasan University
14. Weili Z (1997) Singular perturbations for nonlinear Robin problems. J Comput Appl Math 81:59–74

Key-Dependent Dynamic SBox for KASUMI Block Cipher

Amit Sardar and Dipanwita Roy Chowdhury

Abstract The core strength of a block cipher lies in its nonlinear substitution operation, known as the SBox. However, the presence of static parameters within the
SBox can potentially lead to the exposure of certain information in the ciphertext. In
this paper, we present a methodology for the construction of key-dependent SBoxes.
These key-dependent SBoxes exhibit resistance against linear and differential cryptanalysis. In this paper, we generate key-dependent 7-bit and 9-bit SBoxes for the
KASUMI block cipher. Furthermore, we demonstrate their resistance against known
differential fault attacks

Keywords Differential unity · Nonlinearity · Fixed point · SBox

1 Introduction

Mobile computing devices are experiencing a constant surge in usage, driven by
diverse activities like daily online transactions, video conferencing, etc. Due to
the inherent nature of these digital communications, a strong emphasis on security becomes essential. In response to this demand, 3rd Generation Partnership
Project (3GPP)-based technologies have continuously evolved across generations
of commercial cellular or mobile systems. In 3GPP technology, 64-bit block cipher
KASUMI [1] is mainly used for data confidentiality and integrity purposes. Thus,
the security in 3GPP technology relies solely on the robustness of the KASUMI
cipher. In recent past, several attacks on various versions of KASUMI have been
proposed, employing diverse techniques [2–5]. In the artwork [2], Dunkelman et al.
present a practical key recovery attack on KASUMI. Remarkably, it only requires
2^{26} plaintexts, 2^{30} bytes of memory, and 2^{32} units of time to recover the complete
128-bit key. In article [3], researchers presented a demonstration of an impossible
differential attack. Additionally, Blunden et al. showcased a related key differential

A. Sardar (✉) · D. R. Chowdhury
Indian Institute of Technology, Kharagpur, India
e-mail: amitdare@gmail.com

© The Author(s), under exclusive license to Springer Nature Singapore Pte Ltd. 2024   191
D. Giri et al. (eds.), *Proceedings of the Tenth International Conference on Mathematics
and Computing*, Lecture Notes in Networks and Systems 963,
https://doi.org/10.1007/978-981-97-2069-9_14

attack on a 6-round version of KASUMI in artwork [4]. In articles [5, 6], Gupta et al. demonstrated a differential power analysis attack on the KASUMI cipher. N. Sugio et al. demonstrated integral cryptanalysis on a reduced round KASUMI in article [11]. In reference [10], Jongsung Kim et al. showcased boomerang and rectangle attacks on the KASUMI cipher. Differential fault attack is another significant type of attack within the class of differential cryptanalysis. The first demonstration of this attack was carried out by Boneh et al. in their paper [7]. Subsequently, researchers have applied this class of attack to various block ciphers. Kitae Jeong et al. showcased the initial instance of a differential fault attack on the KASUMI block cipher in their article [8]. Notably, this attack only requires one fault injection and approximately $2^{45.44}$ encryptions to recover a session key on reduced round KASUMI successfully. Another type of differential fault attack was introduced in artwork [9] by Zongyue Wang et al. This methodology requires 2^{32} encryptions and 2^{17} bytes of memory.

Our contribution: Zongyue Wang et al. demonstrated an attack methodology in artwork [9] without proposing any counter-measure techniques. However, this paper introduces a novel counter-measure based on dynamic SBoxes that effectively resists this type of attack. Moreover, we did not come across any articles related to key-dependent dynamic SBoxes for KASUMI. As a result, our work represents the first attempt in this direction.

SBox construction methods that employ chaotic maps or other pseudorandom generators, as described in [14, 15], are typically efficient but often result in SBoxes characterized by relatively low average nonlinearity. On the other hand, algebraic methods, as described in [16, 17] and [18], utilize efficient algebraic constructions to create SBoxes with high nonlinearity. However, these methods have limitations in generating a wide variety of SBoxes, whereas our approach produces a large number of high nonlinearity SBoxes.

Rest of the paper is structured as follows: Section 2 provides a concise introduction to the KASUMI cipher. In Sect. 3, the attack methodology presented in artwork [9] is detailed. Section 4 outlines our proposed counter-measure methodology and its effectiveness against the attack procedure described in the artwork [9]. Finally, in Sect. 5, we conclude and discuss the future scope of our work.

2   KASUMI Block Cipher

KASUMI is a block cipher that is widely used in the telecommunications industry, particularly in the security protocols of mobile networks such as 3G and 4G (UMTS and LTE). It was developed by the 3rd Generation Partnership Project (3GPP) to provide encryption and authentication functions in cellular networks. KASUMI is an eight round feistel structure-based block cipher. Using a 128-bit key, KASUMI conducts the encryption process on a 64-bit block of plaintext. During encryption the plaintext is partitioned into two separate 32-bit strings, denoted as L_0 and R_0, respectively. After that for each integer i satisfying $1 \leq i \leq 8$, the following transformations are executed:

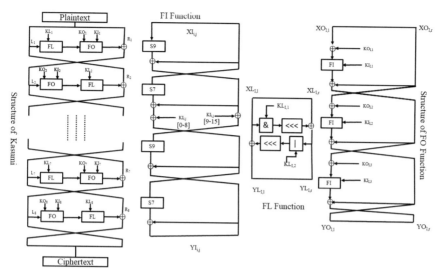

Fig. 1 KASUMI cipher and its architecture

$$R_i = L_{i-1}$$
$$L_i = R_{i-1} \oplus f(L_{i-1}, RK_i) \tag{1}$$

Terminologies used in Eq. 1 are defined as follows:

- R_i represents the value of the right 32-bit string at iteration i.
- L_i represents the value of the left 32-bit string at iteration i.
- R_{i-1} and L_{i-1} are the right and left values from the previous iteration.
- $f(L_{i-1}, RK_i)$ is a function that combines the left value from the previous iteration with a round key RK_i.
- $\oplus$ denotes the bit-wise XOR operation.

The function f at ith round, denoted by $f_i(L_i, RK_i)$, takes a 32-bit input L_i and produces a 32-bit output O using a round key RK_i. This round key consists of the subkey triplet (KL_i, KO_i, KI_i). The $f_i()$ function is constructed from two sub-functions: FL and FO, with associated subkeys KL_i (used with FL) and subkeys KO_i and KI_i (used with FO). The order of execution of FL and FO differs depending on whether it's an even round or an odd round. During even rounds, the execution sequence begins with FO, followed by FL. Conversely, the order is reversed in odd rounds.

Figure 1 depicts the architecture of KASUMI cipher and its component functions.

2.1 FO Function

FO function takes 32-bit XO as data input, along with two sets of 48-bit subkeys: KO_i and KI_i. The 32-bit data input XO is divided into two halves, namely, L_0 and R_0. The 48-bit subkeys are further broken down into three 16-bit subkeys: $KO_i = KO_{i,1} \parallel KO_{i,2} \parallel KO_{i,3}$ and $KI_i = KI_{i,1} \parallel KI_{i,2} \parallel KI_{i,3}$.

For each integer j where $1 \leq j \leq 3$, the following equation need to apply:

$$L_j = R_{j-1}$$
$$R_j = R_{j-1} \oplus FI(L_{j-1} \oplus KO_i, j, KI_{i,j}) \tag{2}$$

Finally, the function yields the 32-bit result denoted as YO ($YO_{I,l} \parallel YO_{I,r}$).

2.2 FI Function

Function FI operates on a 16-bit data input as depicted in Fig. 1 as well as a 16-bit subkey denoted as $KI_{i,j}$. Let's assume the input is M. The input M undergoes a division into two dissimilar portions: 9-bit left half designated as L_0 and 7-bit right half termed as R_0. Moreover, the key $KI_{i,j}$ is partitioned into two components: a 7-bit segment referred to as $KI_{i,j,1}$ and a 9-bit segment named as $KI_{i,j,2}$ and $KI_{i,j}$ is constructed by combining $KI_{i,j,1}$ and $KI_{i,j,2}$. FI function also has two SBoxes: S7, which transforms a 7-bit input to a 7-bit output, and S9, which transforms a 9-bit input to a 9-bit output. Additionally, FI function incorporates two auxiliary functions named as ZE() and TR().

- ZE(x) transforms a 7-bit value x into a 9-bit value by appending two zero bits at the most significant end.
- TR(x) converts a 9-bit value x into a 7-bit value by disregarding the two most significant bits.

Following sequence of actions are required to be executed within the FI function:

1. $L_1 = R_0$
2. $R_1 = S9[L_0] \oplus ZE(R_0)$
3. $L_2 = R_1 \oplus KI_{i,j,2}$
4. $R_2 = S7[L_1] \oplus TR(R_1) \oplus KI_{i,j,1}$
5. $L_3 = R_2$
6. $R_3 = S9[L_2] \oplus ZE(R_2)$
7. $L_4 = S7[L_3] \oplus TR(R_3)$
8. $R_4 = R_3$

At the end FI function yields the 16-bit result as output.

2.3 FL Function

FL function takes a 32-bit data input denoted as XL, along with a 32-bit subkey KL_i. This subkey undergoes a division into two 16-bit subkeys: $KL_{i,1}$ and $KL_{i,2}$. In parallel, the input data XO is partitioned into two equal 16-bit halves referred to as L and R. Consequently, the 32-bit result is given as the concatenation of L' followed by R'. R' is determined by XORing the values of R and the result of a bit-wise cyclic left shift (ROL) applied to the AND operation of L and $KL_{i,1}$. L' is calculated by XORing the values of L and the outcome of a bit-wise cyclic left shift performed on the OR operation of R' and $KL_{i,2}$.

2.4 Key Schedule

KASUMI utilizes a 128-bit key, represented as K. Prior to computation of the round keys, a pair of 16-bit arrays, namely, K_j and K'_j (where j ranges from 1 to 8), are generated according to the following procedure:

- $K = K_1 \parallel K_2 \parallel \cdots \parallel K_8$
- $K'_j = K_j \oplus C_j$

where C_j is the specific constant value outlined in Table 1. Subsequently, the round subkeys are deduced from Kj and K'_j, adhering to the methodology outlined in Table 2.

Table 1 Constants

	Constant
C_1	0x0123
C_2	0x4567
C_3	0x89AB
C_4	0xCDEF
C_5	0xFEDC
C_6	0xBA98
C_7	0x7654
C_8	0x3210

Table 2 Key schedule table

Key	Round1	Round2	Round3	Round4	Round5	Round6	Round7	Round8
$KL_{i,1}$	$K_1 \lll 1$	$K_2 \lll$	$K_3 \lll 1$	$K_4 \lll 1$	$K_5 \lll 1$	$K_6 \lll 1$	$K_7 \lll 1$	$K_8 \lll 1$
$KL_{i,2}$	K_3'	K_4'	K_5'	K_6'	K_7'	K_8'	K_1'	K_2'
$KO_{i,1}$	$K_2 \lll 5$	$K_3 \lll 5$	$K_4 \lll 5$	$K_5 \lll 5$	$K_6 \lll 5$	$K_7 \lll 5$	$K_8 \lll 5$	$K_1 \lll 5$
$KO_{i,2}$	$K_6 \lll 8$	$K_7 \lll 8$	$K_8 \lll 8$	$K_1 \lll 8$	$K_2 \lll 8$	$K_3 \lll 8$	$K_4 \lll 8$	$K_5 \lll 8$
$KO_{i,3}$	$K_7 \lll 13$	$K_8 \lll 13$	$K_1 \lll 13$	$K_2 \lll 13$	$K_3 \lll 13$	$K_4 \lll 13$	$K_5 \lll 13$	$K_6 \lll 13$
$KI_{i,1}$	K_5'	K_6'	K_7'	K_8'	K_1'	K_2'	K_3'	K_4'
$KI_{i,2}$	K_4'	K_5'	K_6'	K_7'	K_8'	K_1'	K_2'	K_3'
$KI_{i,3}$	K_8'	K_1'	K_2'	K_3'	K_4'	K_5'	K_6'	K_7'

3   Differential Fault Attack on KASUMI

In reference [9], the author presents a differential fault attack on the KASUMI cipher, based on specific mathematical insights into the FL and FO functions. Furthermore, the study establishes that only 16-bit word fault is capable of successful key recovery through 2^{32} encryptions. In this experiment, author utilizes eight-word key as follows: $(k_1, k_2, k_3, k_4, k_1, k_2, k_3, k_4)$.

3.1   Fault Model

The central concept involves diminishing the range of potential key candidates through fault injection. The fundamental computational elements of KASUMI are FI and FL functions which operate on 16-bit input data. In accordance with the study detailed in reference [9], the assumption is made that an attacker can be able to introduce a 16-bit word fault to a designated state. To be more precise, by injecting a 16-bit word fault into the output of the penultimate round and skillfully exploiting both accurate and erroneous ciphertexts, attacker can be able to reduce the search space. This methodology effectively truncates the feasible key space from 2^{64} to 2^{32}. Detailed fault propagation architecture is depicted in Fig. 2. The inputs of the final round are designated as L and R, while the ciphertexts are represented by C_L and C_R. The inputs for FI_i and FL are referred to as X_{FI_i} and X_{FL}. The associated outputs are denoted as Y_{FI_i} and Y_{FL}. Additionally, M, N, and O denote the intermediate states as illustrated in Fig. 2. Here, $\triangle$ is the difference between accurate and erroneous state values.

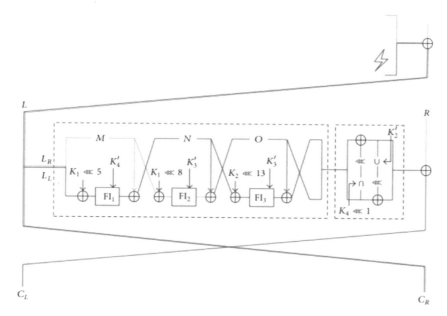

Fig. 2 Differential fault attack on KASUMI

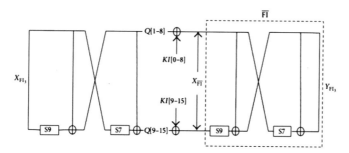

Fig. 3 FI3 of last round

3.2 Attack Procedure

The attack procedures are outlined as follows:

1. Attack procedure commences by selecting a plaintext at random and generating the associated ciphertext using the secret key. For the same plaintext, inject a 16-bit word fault to the position as shown in Fig. 2, so that the left 16-bit word L is corrupted and retain the resulting faulty ciphertext. Noting that, the fault value $(L = C_R)$ is known.
2. After introducing the fault, the value of R remains unchanged. Therefore, we have $\triangle R = 0$ and $\triangle Y_{FL} = \triangle C_L$. Let X and X' represent l bit values, and

consider their difference $\triangle X = X \oplus X'$. Consequently, two distinct properties arise concerning the AND and OR operations such as

$$(X \wedge K) \oplus (X' \wedge K) = \triangle X \wedge K$$
$$(X \vee K) \oplus (X' \vee K) = \triangle X \oplus (\triangle X \vee K) \tag{3}$$

Based on the definition of the FL function and the previous observation 3, we are able to deduce the following insights:

For the known output difference $\triangle Y = \triangle Y_l || \triangle Y_r$ and the FL function's key value (KL_1, KL_2), we can compute the corresponding input difference as follows:

$$\triangle X_l = ((\triangle X_r \oplus (\triangle X_r \wedge KL_2) \lll 1)) \oplus \triangle Y_l$$
$$\triangle X_r = (\triangle X_l \wedge KL_1) \lll 1)) \oplus \triangle Y_r \tag{4}$$

In accordance with Eq. 4, for any assumed values of K_2 and K_4, we are capable of deducing the values of $\triangle X_{FL}$ and $\triangle O$. Due to the corruption affecting solely the left 16-bit word of L, the value of $\triangle M$ is zero. Moreover, the input and output differences of FI_2 are zero. Therefore, $\triangle Y_{FI_1}$ is as follows:

$$\triangle Y_{FI_1} = \triangle M \oplus \triangle N = \triangle M \oplus \triangle Y_{FI_2} \oplus \triangle O = \triangle O \tag{5}$$

Based on Eq. 5, we can say that $\triangle Y_{FI_1}$ depends on K_2 and K_4.

3. In Step 2, K_4 has been guessed. As a result, input and output differences for FI_1, along with KI_1, are determined. Given an input difference $\triangle X$ and an output difference $\triangle Y$ under a random key KI of the FI function, there exist potential input and output values. On average, only one input value can be discovered under KI. According to earlier observation, an average-matching value X_{FI_1} exists that aligns with both the input and output differences. Because $L = C_R$ and $(K_1 \lll 5) = L \oplus FI_1$, we are able to determine the potential values of $k1$.

4. From the earlier steps both accurate and compromised inputs of FI_3 are identified. As illustrated in Fig. 3, only the input value of FI_3 impacts Q. Hence, Q value is deduced. The output difference of $\overline{FI}$ is known through $\triangle X_{FL}$, which was computed in Step 2. Hence, KI_3 (k3) is determined as $KI_3 = X_{\overline{FI}} \oplus Q$.

5. Encrypt the plaintext using the key derived from the preceding steps and verify its correctness. If the key is incorrect, return to Step 2 with an alternative assumption for K2 and K4.

4 Counter-Measure

Here, we can observe that the attack becomes feasible due to the presence of a static SBox. By substituting the static SBox with a key-dependent dynamic SBox, the complexity will increase, potentially reaching 2^{64} once again.

Table 3 SBox

x	S(x)	b3	b2	b1	b0
0	12	1	1	0	0
1	5	0	1	0	1
2	6	0	1	1	0
3	11	1	0	1	1
4	9	1	0	0	1
5	0	0	0	0	0
6	10	1	0	1	0
7	13	1	1	0	1
8	3	0	0	1	1
9	14	1	1	1	0
10	15	1	1	1	1
11	8	1	0	0	0
12	4	0	1	0	0
13	7	0	1	1	1
14	1	0	0	0	1
15	2	0	0	1	0

4.1 Dynamic Sbox Creation

During the creation of dynamic SBox, our objective is not to diminish any of the cryptographic properties from the previous version. To achieve this, we employed a column shuffling technique. Initially, we organize the SBox output on a bit-level basis. Subsequently, each bit vector is treated as a column (e.g., b_3 in Table 3). Using the key value, we then permute these bit vectors with one another according to a predefined rule.

Let's take an example of a 4-bit SBox as defined in Table 3. In this example, b3, b2, b1, and b0 are treated as columns denoted by A, B, C, and D. A 4-bit key, represented as K, is further divided into K3, K2, K1, and K0. The permutation rule for this 4-bit SBox is outlined as follows:

– Permutation order for the columns starts from the left if K3 is zero; otherwise, it starts from the right.
– If K3 is zero and K0 is one, then the permutation occurs in the forward direction with the first neighbor. For instance, in the case of {A,B,C,D}, D's first forward neighbor is C. Therefore, column D is interchanged with column C.
– If K3 is zero and K1 is one, the permutation takes place in the cyclic reverse direction with the second neighbor. Referring to the previous example, column C would be interchanged with column A.

Table 4 SBox

Key	b3	b2	b1	b0
0000	A	B	C	D
0001	A	B	D	C
0010	C	B	A	D
0011	D	B	A	C
0100	A	C	B	D
0101	A	D	B	C
0110	C	A	B	D
0111	D	A	B	C
1000	B	A	C	D
1001	D	A	C	B
1010	B	C	A	D
1011	D	C	A	B
1100	B	D	C	A
1101	A	D	C	B
1110	B	C	D	A
1111	A	C	D	B

- When K3 is zero and K2 is one, the permutation occurs in the cyclic forward direction with the third neighbor. In the context of the previous example, column B would be swapped with column C.
- If K3 is one, the permutation process begins from the right. It involves swapping with the first neighbor in the reverse direction. Referring to the previous example, column A is interchanged with column B.
- If K3 is one and K2 is one, the permutation occurs in the cyclic forward direction with the second neighbor. In the context of the previous example, column B would be interchanged with column D.
- When K3 is one and K1 is one, the permutation happens in the cyclic reverse direction with the third neighbor. Referring to the previous example, column C would be swapped with column B.
- When K3 is one and K0 is one, the permutation takes place in the cyclic forward direction with the third neighbor. In the context of the previous example, column D would be interchanged with column A.

Table 4 provides a comprehensive example of column permutations for the previously mentioned 4-bit SBox, considering all possible key values. In this table, b3, b2, b1, and b0 are denoted as columns A, B, C, and D.

For each distinct key, the permuted column representation is unique. This 4-bit column permutation representation serves as the foundation for all higher bit SBoxes. In the case of a 3-bit SBox, the maximum number of possible permutations is six, whereas there are eight possible key values. Therefore, it's not possible to find distinct permutations for all key values. In the context of KASUMI SBoxes, there are both 7-bit and 9-bit variants. In the case of the 5-bit SBox, a new column E is added to the far right of the existing 4-bit SBox. This addition occurs when K4 equals zero. Conversely, when K4 equals one, column E is permuted with the rightmost 4th bit column. In a similar manner, this process continues incrementally, adding one bit at a time, ultimately generating the 7-bit and 9-bit SBox for KASUMI.

4.2 Sbox Property

By following the aforementioned procedure, we have successfully developed key-dependent 7-bit and 9-bit SBoxes for the KASUMI cipher. The total number of possible 7-bit SBoxes as well as 9-bit SBoxes based on the keys are 128 and 512, respectively. A good SBox should possess high nonlinearity, low differential uniformity, balancedness, low fixed point, and a high algebraic degree property. High non-linearity enhances resistance against linear approximations and differential attacks, while low differential uniformity minimizes possible output differences for a given input difference. Balancedness ensures even output value distribution, and high algebraic degree resists algebraic attacks by complex equations. Low fixed points reduce the chance of self-mapping input values, boosting resistance against algebraic and differential attacks. Tables 5 and 6 provide a comparison of properties between the existing KASUMI SBox and our dynamic SBox (only 5) for 7-bit and 9-bit variants, respectively. In Tables 5 and 6, when all the key bit values are set to zero, it signifies the presence of the existing KASUMI SBox.

In our research, we came across 68 7-bit SBoxes that excel in terms of fixed points when compared to the existing KASUMI 7-bit SBox and 366 distinct 9-bit SBoxes that exhibit enhanced performance in the same aspect. These findings are detailed in [13].

4.3 Complexity Analysis

In step 3 of the attack procedure, a single value X_{FI_1} aligns with both the input and output differences for the existing static SBox. However, if we employ a dynamic SBox in FI_1, earlier mentioned alignment needs to be verified across all SBox possibilities. In FI_1, a 16-bit key is employed in the generation of 7-bit and 9-bit SBoxes. This leads to the requirement of assessing a total of 2^{16} SBoxes. Consequently, the security level is heightened to $\mathcal{O}(2^{16})$. As depicted in Fig. 3, we can deduce k_3 using X_{FI_3} and Y_{FI_3}. If we replace the static SBox with a dynamic one in this context, it would further increase the complexity to an order of 2^{16}. As a result, the overall complexity escalates to 2^{32}. In the above-mentioned attack procedure, the necessity to guess both k_2 and k_4 leads to the overall complexity reaching an equivalent level of a brute force attack, which is 2^{64}. To counter the vulnerability mentioned in article

Table 5 7-bit SBox

Key	Nonlinearity	Differential unity	Balencedness	Algebraic degree	No. of fixed point
0000000	56	2	Yes	3	1
0000001	56	2	Yes	3	0
0000010	56	2	Yes	3	0
0000011	56	2	Yes	3	0
0000100	56	2	Yes	3	0
0000101	56	2	Yes	3	0

Table 6 9-bit SBox

Key	Nonlinearity	Differential unity	Balencedness	Algebraic degree	No. of fixed point
000000000	240	2	Yes	2	2
000000001	240	2	Yes	2	0
000000010	240	2	Yes	2	0
000000011	240	2	Yes	2	0
000000100	240	2	Yes	2	0
000000101	240	2	Yes	2	0

[9], we should implement dynamic SBoxes in FI_1, FI_2, and FI_3 using k_1, k_2, and k_3, respectively.

As a security measure, we conducted an avalanche effect test on both the static and dynamic SBox-based Kasumi cipher. In this test, we intentionally modified a single bit in the plaintext and observed its influence on the ciphertext. Similarly, we also changed a single bit in the key and examined its effect on the ciphertext. The corresponding results, depicted in Table 7, show superior performance in the case of dynamic SBoxes.

We conduct a comparison between the software implementations of the static SBox KASUMI cipher and our dynamic SBox KASUMI cipher utilizing the FELICS [12] framework. Table 8 presents the comparison across three distinct architectures.

Table 7 Avalanche effect on Kasumi with static and dynamic SBox

Plaintext of 64-bit	Key of 128-bit	Bit position changed	Ciphertext of 64-bit	Avalanche effect
ab cd ef 98 76 54 32 1f (static SBox)	01 23 45 67 89 ab cd ef fe dc ba 98 76 54 32 10	Original Plaintext	43 a7 5f c2 86 bf 33 6d	
ab cd ef 98 76 54 32 1e (static SBox)	01 23 45 67 89 ab cd ef fe dc ba 98 76 54 32 10	Plaintext lsb changed	43 01 19 ef 8c d3 bb 17	24 bit
2b cd ef 98 76 54 32 1f (static SBox)	01 23 45 67 89 ab cd ef fe dc ba 98 76 54 32 10	Plaintext msb changed	f8 b5 c9 b0 a0 da 24 87	32 bit
ab cd ef 98 76 54 32 1f (static SBox)	01 23 45 67 89 ab cd ef fe dc ba 98 76 54 32 11	Key lsb changed	73 14 85 a2 01 d1 fe e3	32 bit
ab cd ef 98 76 54 32 1f (static SBox)	81 23 45 67 89 ab cd ef fe dc ba 98 76 54 32 10	Key msb changed	a5 0e 3c a1 9d 62 b0 26	34 bit
ab cd ef 98 76 54 32 1f (dynamic SBox)	01 23 45 67 89 ab cd ef fe dc ba 98 76 54 32 10	Original Plaintext	26 43 95 6d cd 8b ba 01	
ab cd ef 98 76 54 32 1e (dynamic SBox)	01 23 45 67 89 ab cd ef fe dc ba 98 76 54 32 10	plaintext lsb changed	7f 16 70 f3 ef e6 ec 67	33 bit
2b cd ef 98 76 54 32 1f (dynamic SBox)	01 23 45 67 89 ab cd ef fe dc ba 98 76 54 32 10	Plaintext msb changed	c0 cb 6d e9 96 837ab	32 bit
ab cd ef 98 76 54 32 1f (dynamic SBox)	01 23 45 67 89 ab cd ef fe dc ba 98 76 54 32 11	Key lsb changed	93 fa 38 22 fe 38 1e d5	36 bit
ab cd ef 98 76 54 32 1f (dynamic SBox)	01 23 45 67 89 ab cd ef 8e dc ba 98 76 54 32 10	Key msb changed	40 e9 6c c3 1d 97 92 50	30 bit

Table 8 Performance comparison across different architectures

Architecture	SBox type	Code size (byte)	CPU cycle
8-bit AVR ATmega128	Static	3436	2782
8-bit AVR ATmega128	Dynamic	3872	2836
16-bit MSP430F1611	Static	3652	2812
16-bit MSP430F1611	Dynamic	4086	2904
32-bit ARM Cortex-M3	Static	3944	2872
32-bit ARM Cortex-M3	Dynamic	4448	2986

5 Conclusion

Dynamic SBox creates resistance against differential fault attack mentioned in paper [9]. By addressing the concern of fixed-structure SBoxes, our approach yields a more secure ciphertext. In many cases, our dynamic SBoxes exhibit improved properties, particularly in relation to fixed points. This type of SBox construction can be applied to any block cipher that includes SBoxes with a length of 4 bits or more. By the efficient hardware (FPGA) implementation of the proposed methodology, we can enhance the security and throughput.

References

1. TS, ETSI. :135 202 V7. 0.0: Universal Mobile Telecommunications System (UMTS); 690 Specification of the 3GPP confidentiality and integrity algorithms,journal,Vol. 2,Pages:691
2. Dunkelman O, Keller N, Shamir A (2010) A practical-time related-key attack on the KASUMI cryptosystem used in GSM and 3G telephony. In: Advances in cryptology-CRYPTO 2010: 30th annual cryptology conference, August 15–19. Santa Barbara, CA, USA
3. Jia K et al (2012) Improved cryptanalysis of the block cipher KASUMI. In: Selected areas in cryptography: 19th international conference, August 15-16. SAC 2012, Windsor, ON, Canada
4. Blunden M, Escott A (2001) Related key attacks on reduced round KASUMI. In: Fast software encryption: 8th international workshop, April 2–4, 2001. FSE Yokohama, Japan. Revised Papers 8. Springer, Berlin Heidelberg, p 2002
5. Gupta D, Tripathy S, Mazumdar B (2018) Correlation power analysis on KASUMI: attack and countermeasure. In: Security, privacy, and applied cryptography engineering: 8th international conference, SPACE 2018, December 15-19, 2018. Kanpur, India. Proceedings 8. Springer International Publishing
6. Gupta D, Tripathy S, Mazumdar B (2020) Correlation power analysis of KASUMI and power resilience analysis of some equivalence classes of KASUMI S-boxes. J Hardware Syst Secur 4:297–313
7. Boneh D, DeMillo RA, Lipton RJ (1997) On the importance of checking cryptographic protocols for faults. In: International conference on the theory and applications of cryptographic techniques. Springer Berlin Heidelberg, Berlin, Heidelberg

8. Jeong K et al (2011) Fault injection attack on A5/3. In: 2011 IEEE ninth international symposium on parallel and distributed processing with applications. IEEE

9. Wang Z et al (2014) Differential fault attack on KASUMI cipher used in GSM telephony. Mathematical problems in engineering 2014

10. Kim J, Hong S, Preneel B, Biham E, Dunkelman O, Keller N (2012) Related-key boomerang and rectangle attacks: theory and experimental analysis. IEEE Trans Inf Theory 58(7):4948–4966

11. Sugio N, Igarashi Y, Hongo S (2022) Integral cryptanalysis on reduced-round KASUMI. IEICE Trans Fund Electron, Commun Comput Sci 105(9):1309–1316

12. Le Gouguec K, Huynh P (2019) Felics-ae: a framework to benchmark lightweight authenticated block ciphers. In: Proceedings of the 2019 NIST lightweight cryptography workshop

13. https://github.com/amitdare/Kasumi-Sbox

14. Ibrahim S, Alhumyani H, Masud M, Alshamrani SS, Cheikhrouhou O, Muhammad G, Hossain MS, Abbas AM (2020) Framework for efficient medical image encryption using dynamic S-boxes and chaotic maps. IEEE Access 8:160433–160449

15. Zahid AH, Arshad MJ (2019) An innovative design of substitution-boxes using cubic polynomial mapping. Symmetry 11(3), Art. no. 437

16. Ahmad Musheer, Al-Solami Eesa, Alghamdi Ahmed Mohammed, Yousaf Muhammad Awais (2020) Bijective S-boxes method using improved chaotic map-based heuristic search and algebraic group structures. IEEE Access 8:110397–110411

17. Malik MSM, Ali MA, Khan MA, Ehatisham-Ul-Haq M, Shah SNM, Rehman M, Ahmad W (2020) Generation of highly nonlinear and dynamic AES substitution-boxes (S-boxes) using chaos-based rotational matrices. IEEE Access 8:35682–35695

18. Razaq Abdul, Alolaiyan Hanan, Ahmad Musheer, Yousaf Muhammad Awais, Shuaib Umer, Aslam Waqar, Alawida Moatsum (2020) A novel method for generation of strong substitution-boxes based on coset graphs and symmetric groups. IEEE Access 8:75473–75490

Concentric Ellipse Fitting Problem: Theory and Numerical Implementations

Ali Al-Sharadqah⦿, Giuliano Piga, and Ola Nusierat⦿

Abstract The problem of fitting ellipses has been popular since the 1970s, and remains a prominent area of research in statistics, computer vision, and engineering. This paper aims to address the problem of fitting concentric ellipses under general assumptions which started paying more attention recently due to its applications in engineering. We study two methods of obtaining an estimator of the concentric ellipse parameters under this model, namely, the *least squares* (LS) and the *gradient weighted algebraic fits* (GRAF). We address some practical issues in obtaining these estimators. Since our model is nonlinear, obtaining an estimate for the concentric ellipse parameters requires the implementation of numerical minimization schemes. We propose and compare several minimization schemes, and provide several initial guesses which yield the best convergence rates.

Keywords Concentric ellipse fitting · Iterative methods · Initial guess

1 Introduction

Computer vision is a broad field with a host of data-centric challenges to be addressed. One such class of challenges, for example, is image recognition problems, in which an image is to be classified as belonging to a class of images or not. Image recognition problems are often handled with generalized methods, such as convolutional neural networks [1]. The methods used to solve these problems are often robust but are not designed for a specific type of data. Data-specific models can be used when more

A. Al-Sharadqah (✉) · G. Piga
California State University Northridge, Northridge, CA 91330, USA
e-mail: asharadqah@pmu.edu.sa

G. Piga
e-mail: giuliano.piga.42@my.csun.edu

A. Al-Sharadqah · O. Nusierat
Prince Mohammad Bin Fahid University, Dhahran, Saudi Arabia
e-mail: onusierat@pmu.edu.sa

© The Author(s), under exclusive license to Springer Nature Singapore Pte Ltd. 2024 207
D. Giri et al. (eds.), *Proceedings of the Tenth International Conference on Mathematics and Computing*, Lecture Notes in Networks and Systems 963,
https://doi.org/10.1007/978-981-97-2069-9_15

than classification is of interest. For example, if an object is known (or assumed) to belong to a certain class of objects, one might be interested in its dimensions, or some parameters of its underlying shape, given an image of it. These sorts of problems are handled with another area of computer vision known as *Geometric Estimation*. In this paper, we will discuss a special problem in geometric fitting, namely, concentric ellipse fitting. This has been an active research area recently because of its applications in camera calibration [7], and biometrics, specifically, iris recognition [13]. Moreover, it has tremendous applications in astronomy [10], robotic biometrics [2], medicine [3], and archaeology and engineering [6].

Ellipses are a commonly occurring shape in nature. Moreover, the projection of circular objects into the image plane is elliptical. This has made the problem of ellipse fitting popular among researchers and engineers. The general methodology for fitting an ellipse is to estimate the parameters of an ellipse. Several methods, such as the *Maximum Likelihood Estimation* method (MLE) or *Orthogonal Distance Regression* (ODR), have proven to provide the best estimates of the ellipse parameters. These methods, however, pose a major drawback, being the computational time required to obtain those estimates. Alternatively, cheaper but less accurate methods, such as the *algebraic methods*, were developed. Moreover, when data is corrupted by outliers those approaches fail, and as such other approaches such as RANSAC or its variant shall be used [11, 12].

This paper focuses on fitting concentric ellipses. We develop a method that provides accurate estimates of the parameters. Along these lines, Al-Sharadqah and Rulli [2] study the problem of concentric ellipse fitting under some restrictive constraints. Al-Sharadqah and Rulli [2] assume that the tilt angles of the concentric ellipses are equal and also the major and the minor axis lengths of the two ellipses are proportional (i.e., if Ellipse 1 has major and minor axes a_1 and b_1 and Ellipse 2 has major and minor axes a_2 and b_2, then $\frac{a_1}{b_1} = \frac{a_2}{b_2}$.) This allows them to represent the problem algebraically and develop several non-iterative methods. However, imposing those constraints adds limitations to their practical uses in real-life applications. For example, one might be interested in identifying the structure of galaxies [10] by estimating the mass-to-luminosity ratio of nearby galaxies (see Fig. 1). In this case, the Al-Sharadqah and Rulli algorithm is not applicable, and new algorithms are needed to handle this general case. In this paper, we relax the geometric constraints of Al-Sharadqah and Rulli's approach. We will only assume that the data are perturbed digitized pixels coming from concentric ellipses where the "noise" in the data is independent and identically normally distributed. In notations, let $\boldsymbol{m}_{ij}$'s be perturbed points coming from a pair of concentric ellipses, where the point $\boldsymbol{m}_{ij}$ is the jth point of the ith ellipse, and as such $i \in \{1, 2, \ldots, K\}$, $j \in \{1, \ldots, n_i\}$. Without loss of generality, let us assume the number of concentric ellipses is $K = 2$. The extension to $K > 2$ case is rather simple and is omitted here. Let us denote the true point that lies exactly on the ellipse by $\tilde{\boldsymbol{m}}_{ij} = (\tilde{x}_{ij}, \tilde{y}_{ij})$. That is $\tilde{\boldsymbol{m}}_{ij}$ satisfies

$$\mathcal{P}(\boldsymbol{m}_{ij}, \boldsymbol{\theta}) = 0, \tag{1}$$

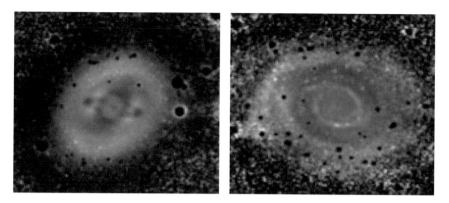

Fig. 1 Galaxy images

where $\mathcal{P}(\boldsymbol{m}_{ij}, \boldsymbol{\theta}) := p_{ij} = \frac{T_{ij}^2}{a_i^2} + \frac{T_{ij}'^2}{b_i^2} - 1$, and

$$
\begin{aligned}
T_{ij} &= \left(x_{ij} - x_c\right)\cos(\psi_i) + \left(y_{ij} - y_c\right)\sin(\psi_i) \\
T_{ij}' &= -\left(x_{ij} - x_c\right)\sin(\psi_i) + \left(y_{ij} - y_c\right)\cos(\psi_i),
\end{aligned}
\tag{2}
$$

where x_c and y_c form the coordinate for the center of the ellipse. The parameters a and b make up the major and minor axis lengths, respectively, and ψ represents the tilt angle as depicted in Fig. 2 (left). However, the true points are unobservable but their proxies $\boldsymbol{m}_{ij}$ are detected. Accordingly, we assume the observed points are true points contaminated by some noise, i.e., $\boldsymbol{m}_{ij} = (x_{ij}, y_{ij})^\top$ are perturbed from their true points as $\boldsymbol{m}_{ij} = \tilde{\boldsymbol{m}}_{ij} + \boldsymbol{n}_{ij}$, where $\boldsymbol{n}_{ij} = \left(\delta_{ij}, \epsilon_{ij}\right)^\top$.

Our contribution to this paper is summarized as follows. First, we develop two estimators for our problem. Each one is obtained by minimizing an objective function. Due to the highly nonlinear nature of our problem, we partially linear the problem, which reduces the complexity of the problem significantly. However, the problem continues to be nonlinear and shall be solved by an iterative approach after seeding the algorithm with a reliable initial guess. Therefore, we propose three initial guesses and we implement three different numerical schemes. Based on our intensive numerical experiments, we recommend which optimizer and initial guess produce the most reliable results by computing the average run time (ART) and the divergence rate. This paper is organized as follows. Section 2 overviews the problem of fitting a single ellipse. In Sect. 3, we discuss our research problem and its numerical implementation including the proposed initial guesses. Section 4 summarizes our numerical experiments on real and synthetic data while Sect. 5 concludes our findings.

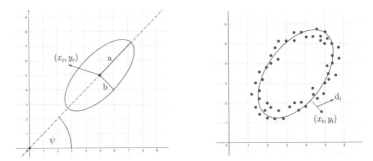

Fig. 2 Ellipse geometry (left) and perturbed data along ellipse (right). d_i represents the Euclidean distance between the observed point and the fitted ellipse

2 Overview of Fitting an Ellipse to Data

Although the method proposed in this paper is for fitting concentric ellipses to data and it must be handled differently than that of the single ellipse fitting problem, we devote this section to discuss briefly the single ellipse case. First, consider the general equation for an ellipse $\mathcal{P}(\boldsymbol{m}, \boldsymbol{\theta}) = 0$ and assume n observed points $\boldsymbol{m}_j = (x_j, y_j)$, $j = 1, \ldots, n$ are recorded. We assume the additive model where $\boldsymbol{m}_j = \tilde{\boldsymbol{m}}_j + \boldsymbol{n}_j$, where the noisy vector $\boldsymbol{n}_j = (\delta_j, \epsilon_j)^\top \sim N(\boldsymbol{0}_2, \sigma^2 \boldsymbol{I}_2)$, where $\boldsymbol{I}_2$ is identity matrix of size 2. According to those assumptions, one can estimate the parameter using the MLE of $\tilde{\boldsymbol{\theta}}$ which is equivalent to minimizing $S(\boldsymbol{\theta}) = \sum_{j=1}^{n} \left((x_j - \tilde{x}_j)^2 + (y_j - \tilde{y}_j)^2 \right)$. This minimizing is subtle here because the true point $\tilde{\boldsymbol{m}}_j$ satisfies $\mathcal{P}(\tilde{\boldsymbol{m}}_j, \boldsymbol{\theta}) = 0$. Kanatani [9] proves that this problem is equivalent to minimizing the sum of the squares of the orthogonal distance of $\boldsymbol{m}_j$ from the true curve; i.e., $S(\boldsymbol{\theta}) = \sum_{j=1}^{n} \left((x_j - \tilde{x}_j^*)^2 + (y_j - \tilde{y}_j^*)^2 \right)$, where $\tilde{\boldsymbol{m}}_j^* = (x_j^*, y_j^*)$ represents a point of the true curve such that the (orthogonal) Euclidean distance d_j of $\boldsymbol{m}_i$ to the true curve is the smallest as shown in Fig. 2 (right). However, the orthogonal distance regression that minimizes the sum of the squares of the orthogonal distances turns out to be the solution of a polynomial of degree 4, which is numerically unstable, and as such, it shall be avoided. This is just a subproblem that needs to be solved as one of several other steps in the minimization process. On the other hand, one can obtain an estimate by minimizing the sum of the squares of algebraic distances, and as such, it is referred in literature by algebraic fits. The algebraic approach provides much faster estimates of the ellipse parameters. The simplest approach to estimate the parameters of (3) is the Least Squares (LS) approach. In the LS, the following is minimized $\mathcal{J}(\mathcal{A}) = \sum_{j=1}^{n} (\boldsymbol{\xi}_j^\top \mathcal{A})^2$, where $\boldsymbol{\xi}_j = (x_j^2, 2x_j y_j, y_j^2, x_j, y_j, 1)^\top$ and $\mathcal{A} = (A, B, C, D, E, F)^\top$. If we define $\boldsymbol{M} = \sum_{j=1}^{n} \boldsymbol{\xi}_j \boldsymbol{\xi}_j^\top$, we can further simplify $\mathcal{J}$ to become $\mathcal{J}(\mathcal{A}) = \mathcal{A}^\top \boldsymbol{M} \mathcal{A}$. The advantage of using this algebraic representation is that it provides a closed-form solution in terms of $\mathcal{A}$ depending on the constraints. It is worth mentioning here that the parametric space of the conic is represented by

$$Ax_j^2 + 2Bx_jy_j + Cy_j^2 + 2Dx_j + 2Ey_j + F = 0, \tag{3}$$

which includes an ellipse, hyperbola, parabola, parallel lines, imaginary lines, and many other shapes, depending on the values of the parameters. For example, if $AC - B^2 = 0$, then (3) will be the expression for a parabola, but if $AC - B^2 < 0$, the expression is that of a hyperbola. Only if $AC - B^2 > 0$ will the expression be that of an ellipse. Therefore, a constraint shall be imposed on the parametric space to remove any indeterminacy. To avoid this, a constraint, for example, $\|\mathcal{A}\|^2 = 1$, is imposed. This is a constrained minimizing problem. Differentiating $\mathcal{A}^\mathsf{T} M\mathcal{A} - \lambda \left(\|\mathcal{A}\|^2 - 1 \right)$ with respect to $\mathcal{A}$ and setting the result equal to zero yields $M\hat{\mathcal{A}}_\mathrm{L} = \lambda\hat{\mathcal{A}}_\mathrm{L}$, which is an eigenvalue problem.

The LS estimator is fast and easy to compute but it suffers from heavy bias. Consequently, other algebraic-based methods have been proposed by imposing different constraints. These constraints are typically of the form $\mathcal{A}^\mathsf{T} N\mathcal{A} = 1$. A common method for fitting ellipses is called the Taubin method. The Taubin method imposes a different constraint that takes the form $N_\mathrm{T} = 4\sum_{j=1}^{n} V_j$, where

$$V_j = \begin{pmatrix} x_j^2 & x_jy_j & 0 & x_j & 0 & 0 \\ x_jy_j & x_j^2 + y_j^2 & x_jy_j & y_j & 0 & 0 \\ 0 & x_jy_j & y_j^2 & 0 & 0 & 0 \\ x_j & y_j & 0 & 1 & 0 & 0 \\ 0 & 0 & 0 & 0 & 0 & 0 \\ 0 & 0 & 0 & 0 & 0 & 0 \end{pmatrix}_{6\times6}.$$

Therefore, the Taubin method solves $M\hat{\mathcal{A}}_\mathrm{T} = \lambda N_\mathrm{T}\hat{\mathcal{A}}_\mathrm{T}$. Since M and N_T are positive semi-definite matrices, λ is non-negative. Thus, the estimator $\hat{\mathcal{A}}_\mathrm{T}$ is the generalized eigenvector corresponding to the smallest possible generalized eigenvalue of the generalized eigenvalue problem. The Taubin method is considered to be a good method for obtaining an estimate for the parameter vector $\mathcal{A}$. Therefore, we will implement the Taubin method in the process of developing a good initial guess for our proposed algorithms.

3 Concentric Ellipses Under General Assumptions

We turn our attention to estimate $\theta = (x_c, y_c, a_1, b_1, \psi_1, a_2, b_2, \psi_2)$. In the problem of fitting all other primitive geometric shapes, such as circles, ellipses, and concentric ellipses (under the Al-Sharadqah and Rulli model), the LS is a non-iterative method that is easy to compute. However, under our general model, the situation is different here. In this case $\mathcal{P}(m_{ij}, \theta)$ is a highly nonlinear function, which renders the minimization process of $\mathcal{F}$ a bit difficult to deal with. The reason here is that it is infeasible to linearize $\mathcal{P}$. As a remedy, we propose to "partially linearize" our problem with a change of variables by defining a new parameter vector ϕ as $\phi = (x_c, y_c, A_1, B_1, \psi_1, A_2, B_2, \psi_2)^\mathsf{T}$, where $A_i = \frac{1}{a_i^2}$, $B_i = \frac{1}{b_i^2}$, $i = 1, 2$. After ϕ is estimated, the geometric parameter vector θ can be retrieved.

3.1 Developed Estimators

In this paper, we will discuss our proposed methods: the Least Squares (LS) method and the Gradient Weighted Algebraic Fit (GRAF). Here we describe them as follows.

1. Least Squares (LS) The LS is the simplest and fastest method and it can be used to obtain an estimator $\hat{\phi}_L$ by minimizing the following objective function:

$$\hat{\phi}_L = \operatorname{argmin}_\phi \mathcal{F}_L(\phi) = \operatorname{argmin}_\phi \sum_{i=1}^{2} \sum_{j=1}^{n_i} p_{ij}^2. \tag{4}$$

In order to minimize (4), one can use gradient methods which require the computation for the gradient of (4). Here

$$\nabla \mathcal{F}_L = \begin{pmatrix} \frac{\partial \mathcal{F}_L}{\partial x_c} \\ \frac{\partial \mathcal{F}_L}{\partial y_c} \\ \frac{\partial \mathcal{F}_L}{\partial A_1} \\ \frac{\partial \mathcal{F}_L}{\partial B_1} \\ \frac{\partial \mathcal{F}_L}{\partial \psi_1} \\ \frac{\partial \mathcal{F}_L}{\partial A_2} \\ \frac{\partial \mathcal{F}_L}{\partial B_2} \\ \frac{\partial \mathcal{F}_L}{\partial \psi_2} \end{pmatrix} = 4 \sum_{i=1}^{2} \sum_{j=1}^{n_i} \begin{pmatrix} p_{ij} \left(-T_{ij} A_i C_i + T'_{ij} B_i S_i \right) \\ p_{ij} \left(-T_{ij} A_i S_i - T'_{ij} B_i C_i \right) \\ \frac{1}{2} p_{1j} T_{1j}^2 \hat{\delta}_{1i} \\ \frac{1}{2} p_{1j} T_{1j}'^2 \hat{\delta}_{1i} \\ p_{1j}(T_{1j} T'_{1j} A_1 - T_{1j} T'_{1j} B_1) \hat{\delta}_{1i} \\ \frac{1}{2} p_{2j} T_{2j}^2 \hat{\delta}_{2i} \\ \frac{1}{2} p_{2j} T_{2j}'^2 \hat{\delta}_{2i} \\ p_{2j}(T_{2j} T'_{2j} A_2 - T_{2j} T'_{2j} B_2) \hat{\delta}_{2i} \end{pmatrix}, \tag{5}$$

where $C_i = \cos(\psi_i)$ and $S_i = \sin(\psi_i)$ for $i = 1, 2$. Also, $\hat{\delta}_{ki}$ stands for the Dirac delta function, i.e., $\hat{\delta}_{ki} = 1$ when $i = k$ and zero otherwise.

2. Gradient Weighted Algebraic Fit (GRAF) The LS provides a good estimator when we fit data coming from large arcs of the ellipses; however, its performance dramatically reduces as the arc length from which the data is obtained becomes shorter. Alternatively, one can implement the weighted objective function. The estimator of this objective function is called the Gradient Weighted Algebraic Fit (GRAF), which is virtually an approximation of the maximum likelihood estimator. The GRAF minimizes the weighted sum of the squares of the algebraic distances p_{ij}. The estimator from GRAF (denoted by $\hat{\phi}_G$) is

$$\hat{\phi}_G = \operatorname{argmin}_\phi \mathcal{F}_G(\phi) = \operatorname{argmin}_\phi \sum_{i=1}^{2} \sum_{j=1}^{n_i} \frac{p_{ij}^2}{\|\nabla p_{ij}\|^2}, \quad \text{where} \nabla p_{ij} = 2 \begin{pmatrix} T_{ij} A_i C_i - T'_{ij} B_i S_i \\ T_{ij} A_i S_i + T'_{ij} B_i C_i \end{pmatrix}, \tag{6}$$

and as such, we define $w_{ij} := \frac{1}{4} \|\nabla p_{ij}\|^2 = (T_{ij}^2 A_i^2 + T_{ij}'^2 B_i^2)$. (Here we used the trigonometric identity $C_i^2 + S_i^2 = 1$.) As in the LS, GRAF can be minimized using gradient methods, which require computing the Jacobian vector of $\mathcal{F}_G$, given by

$$\nabla \mathcal{F}_{\mathrm{G}} = \frac{1}{2} \sum_{i=1}^{2} \sum_{j=1}^{n_i} w_{ij}^{-2} \begin{pmatrix} p_{ij}^2 (T_{ij} A_i^2 C_i - T_{ij}' B_i^2 S_i) - 2 p_{ij} w_{ij} (T_{ij} A_i C_i - T_{ij}' B_i S_i) \\ p_{ij}^2 (T_{ij} A_i^2 S_i + T_{ij}' B_i^2 C_i) - 2 p_{ij} w_{ij} (T_{ij} A_i S_i + T_{ij}' B_i C_i) \\ (p_{1j} w_{1j} T_{1j}^2 - p_{1j}^2 T_{1j}^2 A_1) \hat{\delta}_{1i} \\ (p_{1j} w_{1j} T_{1j}'^2 - p_{1j}^2 T_{1j}'^2 B_1) \hat{\delta}_{1i} \\ (2 p_{1j} w_{1j} T_{1j} T_{1j}' (A_1 - B_1) - p_{1j}^2 T_{1j} T_{1j}' (A_1^2 - B_1^2)) \hat{\delta}_{1i} \\ (p_{2j} w_{2j} T_{2j}^2 - p_{2j}^2 T_{2j}^2 A_2) \hat{\delta}_{2i} \\ (p_{2j} w_{2j} T_{2j}'^2 - p_{2j}^2 T_{2j}'^2 B_2) \hat{\delta}_{2i} \\ (2 p_{2j} w_{2j} T_{2j} T_{2j}' (A_2 - B_2) - p_{2j}^2 T_{2j} T_{2j}' (A_2^2 - B_2^2)) \hat{\delta}_{2i} \end{pmatrix}. \quad (7)$$

3.2 Numerical Schemes

In practice, these methods require numerical optimization techniques; hence, the quality of the estimators will depend on the method as well as the optimization technique we choose. Due to the nonlinearity of our objective functions, numerical optimization techniques shall be implemented. The Newton–Raphson method is a popular minimization method that converges quickly when it converges. However, this method is very sensitive to the seeded initial guess. Therefore, we will implement other more robust iterative approaches. We will implement three methods (BFGS, L-BFGS-B, and Nelder–Mead). Then we will compare them in terms of their divergence rates and the average number of iterations they take to converge (if at all). Now we briefly describe those popular numerical schemes.

I. BFGS. The BFGS algorithm is one of the quasi-Newton methods, where the method of approximating the Hessian matrix is updated with each iteration.

II. L-BFGS-B. L-BFGS-B (Limited memory-BFGS-Box) addresses another computational cost issue involved in the BFGS algorithm, as well as imposing some box constraints on the parameters being estimated.

III. Nelder–Mead. Nelder–Mead is commonly used to optimize functions in high-dimensional space. Nelder–Mead is a much more complex algorithm than the previous two. It involves three steps: 1. Ordering, 2. Centroid, 3. Transformation, where a new working simplex is generated from the current one.

3.3 Initial Guess

Note that all those algorithms depend on the initial guess that is seeded. This is a critical issue in order to improve the convergence rate. Therefore, we will discuss three initial guesses used to initialize those algorithms. Because our objective functions are highly nonlinear, it is important to form an initial guess that is in the proximity of the global minimum. There are several factors to consider here: Using an initial guess that results in the best estimate is the main objective, but we need also to keep in mind the difficulty or cost (run time) in obtaining it. Here we propose three approaches to tackle this problem:

1. Center Averaging. One approach to form an initial guess is to fit each ellipse separately and take an average of their center estimate. Each initial guess we will assess here requires first the fitting of single ellipses via Taubin's method. Simply fitting each ellipse individually and taking an average of the center estimates to form an estimate of the center for the concentric ellipses will serve as an initial guess of its own. That is, we will obtain $\hat{\phi}_0$ by fitting the two ellipses separately and obtaining $\hat{\phi}_i = \left(\hat{x}_{ci}, \hat{y}_{ci}, \hat{A}_i, \hat{B}_i, \hat{\psi}_i \right)^{\top}$, $i = 1, 2$, and define the initial guess by $\hat{\phi}_0 = \left(\frac{\hat{x}_{c1}+\hat{x}_{c2}}{2}, \frac{\hat{y}_{c1}+\hat{y}_{c2}}{2}, \hat{A}_1, \hat{B}_1, \hat{\psi}_1, \hat{A}_2, \hat{B}_2, \hat{\psi}_2 \right)^{\top}$.

2. Stepwise Search. Alternatively, one can update the center estimates based on the value of our objective function. We assume a good initial guess for the center lies on the line segment connecting each center from the individual fits. That is, fit the two ellipses as in the center averaging initial guess to obtain $\hat{\phi}_i^{\top} = (\hat{x}_{ci}, \hat{y}_{ci}, \hat{A}_i, \hat{B}_i, \hat{\psi}_i)$, $i = 1, 2$. Next, evaluate $F_i = \mathcal{F}((\hat{x}_{ci}, \hat{y}_{ci}, \hat{A}_1, \hat{B}_1, \hat{\psi}_1, \hat{A}_2, \hat{B}_2, \hat{\psi}_2))$ for $i = 1, 2$, and choose $\hat{\phi}_0^{\top} = (\hat{x}_{ci}, \hat{y}_{ci}, \hat{A}_1, \hat{B}_1, \hat{\psi}_1, \hat{A}_2, \hat{B}_2, \hat{\psi}_2)$ based on which i produced a smaller value of F_i. For example, if $F_1 < F_2$, choose $(\hat{x}_{c1}, \hat{y}_{c1})$ as the first step in finding an initial guess for the center values. The next step is to obtain m equally spaced points on the line segment connecting $(\hat{x}_{c1}, \hat{y}_{c1})$ and $(\hat{x}_{c2}, \hat{y}_{c2})$, and label these points $(\hat{x}_{ck_1}, \hat{y}_{ck_1}), \ldots, (\hat{x}_{ck_m}, \hat{y}_{ck_m})$. Then, compute $F_{k_1} = \mathcal{F}((\hat{x}_{ck_1}, \hat{y}_{ck_1}, \hat{A}_1, \hat{B}_1, \hat{\psi}_1, \hat{A}_2, \hat{B}_2, \hat{\psi}_2))$. If $F_{k_1} < F_i$, then update $(\hat{x}_{c1}, \hat{y}_{c1})$ to $(\hat{x}_{ck_1}, \hat{y}_{ck_1})$ and repeat the previous step until $F_{k_j} > F_{k_{j+1}}$. Once you reach a value of j such that $F_{k_j} > F_{k_{j+1}}$, choose $(\hat{x}_{ck_j}, \hat{y}_{ck_j})$ as the initial guess for the center.

3. Pooling. Like the other two initial guesses, we fit each ellipse separately to initialization of all parameters besides the center. Here we simply take $\bar{x} = \frac{1}{n} \sum_{i=1}^{2} \sum_{j=1}^{n_i} x_{ij}$ and $\bar{y} = \frac{1}{n} \sum_{i=1}^{2} \sum_{j=1}^{n_i} y_{ij}$ where $n = n_1 + n_2$ as the initial guess for the center. Therefore, we call this initial guess "pooling initial guess".

4 Numerical Experiments

In the previous section, we have discussed three algorithms and three initial guesses but we have not discussed which one shall be used. In this section, we ran several Monte Carlo experiments that helped us shed more light on each method and the choice of initial guess. There are several important factors in determining an optimizer. The divergence rate is important in that we would like to have some level of confidence that our estimate is obtained from finding a minimum in the objective function. An optimizer that diverges often will not be considered feasible. Another important aspect is the run time till convergence. Among optimizers, the one that converges and provides a solution faster is preferable. Of course, there are trade-offs to consider between divergence rate and run time. Finally, we compare the LS and

GRAF in terms of Mean-Squared Error (MSE). All the experiments we ran here use $n_1 = 20$ and $n_2 = 35$ equidistant points positioned on varying ellipse arc lengths with central angle $\rho = 120°$, $180°$, and $240°$. The parameters of the two concentric ellipses were set to $a_1 = 2$, $b_1 = 1$, $\psi_1 = 0$, $a_2 = 3$, $b_2 = 2$, $\psi_2 = \frac{\pi}{4}$. We repeated the experiments 2000 times and used three methods (Nelder–Mead, BFGS, and L-BFGS-B algorithm). Moreover, we consider three initial guesses: (i) center averaging (black colored), (ii) stepwise (red colored), and (iii) pooling (blue colored).

For each configuration, we generated the specified number of true points on the true ellipses described above. We then simulated $n = n_1 + n_2$ noisy values (δ_k, ϵ_k) from a normal distribution with mean zero and standard deviation σ. These observations are then added to the true points. Then three optimization algorithms with various initial guesses were implemented and compared. Finally, it is worth mentioning here that our results summarized here are obtained from the LS only as the results from GRAF agree with our results here.

(I) Investigating initial guesses. In this set of experiments we investigated the three initial guesses. Figures 3 and 4 show the divergence rate and average run time (ART), respectively, for BFGS and Nelder–Mead algorithms. It is worth mentioning that the L-BFGS-B algorithm fails for short arc lengths and it shows similar results as those of BFGS for long arcs. Therefore, we exclude its results here. As shown in Figs. 3 and 4, it becomes clear that all initial guesses perform roughly the same in terms of divergence rate under each optimizer but pooled and center averaging initial guesses yield roughly identical results in terms of ART, while the stepwise initial guess requires the longest ART. To conclude, we recommend the use of the center averaging initial guess.

(II) Investigating optimizers. In this set of experiments, we will focus on the best optimization method. Our numerical results are depicted in Figs. 5 and 6, which show the divergence rate and ART, respectively, of BFGS (solid), Nelder–Mead (dotted), and L-BFGS-B (dashed) algorithms under three scenarios: (i) $\rho = 240°$ (left subfigures), (ii) $\rho = 180°$ (center subfigures), and (iii) $\rho = 120°$ (right subfigures). Figures 5 and 6 show that the BFGS algorithm requires less time than the Nelder–Mead and L-BFGS-B algorithms for each value of σ under (i). In addition, we see that the BFGS and L-BFGS-B algorithms have coinciding divergence rates, being slightly lower than that of the Nelder–Mead algorithm. Under (ii) the figures show that the BFGS algorithm is faster than Nelder–Mead and has a lower divergence rate. Finally, under (iii) one can see that the BFGS algorithm has a higher ART as σ increases but more importantly, it has a lower divergence rate. In summary, one can reliably use the BFGS algorithm to obtain estimates under our model with the center averaging initial guess.

(III) Comparison between LS versus GRAF. In this set of experiments, we compare the performance between the LS and GRAF. In this experiment, we positioned $n_1 = 30$ and $n_2 = 45$ equidistant points that lie on the true ellipses and we considered the following true values of the parameters: $x_c = 0$, $y_c = 0$, $b_1 = 1$, $a1 = 2$, $a_2 = 3$, $b_2 = 2$, $\psi_1 = 0$, $\psi_2 = \frac{\pi}{4}$. The true points were distributed along arcs of circular angles $360°$ (left subfigures), $240°$ (centered subfigures), and $180°$ (right subfigures).

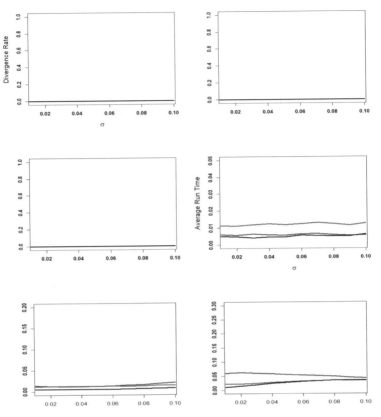

Fig. 3 The divergence rate (top row) and the ART (bottom row) of the BFGS algorithm using the three initial guesses for $\rho = 240°$, $180°$, and $120°$, respectively

In each experiment, $N = 2000$ samples were generated by adding white noise of level σ to the true points and we computed the estimates of the LS and GRAF. Then we computed the *empirical normalized MSE* ($\widehat{\text{NMSE}}$) of ϕ for each method using $\widehat{\text{NMSE}}(\hat{\phi}) = \frac{1}{N\sigma^2} \sum_{i=1}^{N} \|\hat{\phi}_i - \tilde{\phi}\|^2$, where $\hat{\phi}_i$ is the ith estimate of the true parameter vector $\tilde{\phi}$. The dashed line in Fig. 7 represents $\widehat{\text{NMSE}}(\hat{\phi}_L)$, the solid line represents the $\widehat{\text{NMSE}}(\hat{\phi}_G)$, and the dotted line represents the normalized Cramér–Rao lower bound (CRB). Then the empirical NMSEs were plotted against σ. For smaller arc lengths, such as $180°$, the two methods diverge faster, hence we computed the normalized MSE for σ up to .04 only as the figures on the right column reveal. Our experiments show that the GRAF estimator outperforms the LS estimator in terms of the empirical normalized MSE. For small values of σ, $\widehat{\text{NMSE}}(\hat{\phi}_G)$ approaches the CRB regardless of the arc lengths, while $\widehat{\text{NMSE}}(\hat{\phi}_L)$ does not. The differences between the normalized MSEs of the two methods increase as the arc length increases.

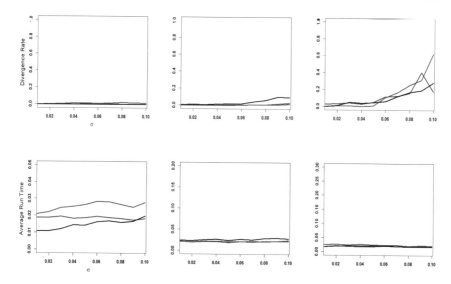

Fig. 4 The divergence rate (top row) and the ART (bottom row) of the Nelder–Mead algorithm using the three initial guesses for $\rho = 240°$, $180°$, and $120°$, respectively

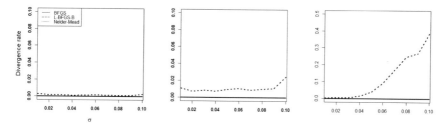

Fig. 5 The divergence rate versus σ for (i) $\rho = 240°$ (left subfigures), (ii) $\rho = 180°$ (center subfigures), and (iii) $\rho = 120°$ (right subfigures), respectively, using the center averaging initial guess. Here BFGS (solid), Nelder–Mead (dotted), and L-BFGS-B (dashed)

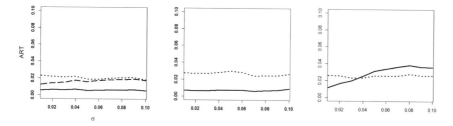

Fig. 6 The ART versus σ for (i) $\rho = 240°$ (left subfigures), (ii) $\rho = 180°$ (center subfigures), and (iii) $\rho = 120°$ (right subfigures), respectively, using the center averaging initial guess. Here BFGS (solid), Nelder–Mead (dotted), and L-BFGS-B (dashed)

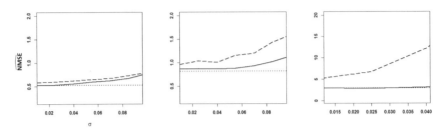

Fig. 7 The empirical normalized MSE for the LS (dashed) and GRAF (solid) versus σ. The normalized Cramér–Rao lower bound is given by the dotted line when data are distributed along arcs with circular angles 360° (left), 240° (center), and 180° (right)

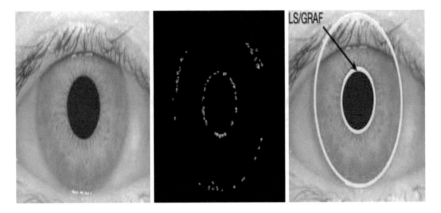

Fig. 8 The image of the eye's iris superimposed by the fitted methods

To corroborate our analytic findings with a set of practical data, we applied our algorithms to real images and then assessed their performances. Since one of the most important applications for fitting concentric ellipses appears in iris recognition [5, 8], we applied all methods to the binary image of a human iris. The *Canny Edge Detection* technique [4] was implemented to extract the digitized data points from the inner and outer concentric ellipses. We have a total of $n_1 = 50$ and $n_2 = 60$ points from both concentric ellipses, respectively. In Fig. 8, the LS and the GRAF methods were superimposed on the image. One can easily see that the LS produces estimates with heavy biases, and as such, their estimates for the actual inner and outer boundaries are less satisfactory. This is not an unexpected observation regarding the LS method which is widely known to be heavily biased. This observation deserves a thorough theoretical investigation. On the other hand, the GRAF provides the closest fit to the inner ring of the iris.

5 Conclusion

In this paper, concentric ellipses fitting problems under general geometric assumptions have been studied and two methods (LS and GRAF) have been developed and implemented using three optimizers. We also propose three initial guesses. Based on our intensive numerical experiments, the superiority of using the BFGS method over other optimizers has been verified. Also, it was found that the center averaging initial guess provides the best initial guess. Moreover, our numerical experiments show the superiority of GRAF over LS.

References

1. Aggarwal CC (2018) Neural networks and deep learning: a textbook. Springer
2. Al-Sharadqah A, Rulli L (2022) New methods for detecting concentric objects with high accuracy. Measurement 188
3. Bankman (2008) Handbook of medical image processing and analysis. Elsevier Science
4. Canny J (1986) A computational approach to edge detection. IEEE Trans Pattern Anal Mach Intell 8:679–698
5. CASIA (2012) Casia iris image database. http://biometrics.idealtest.org/
6. Chernov N (2010) Circular and linear regression: fitting circles and lines by least squares, vol 117. Chapman & Hall/CRC, Boca Raton
7. Kweon IS, Kim JS (2022) A camera calibration method using concentric circles of vision applications. In: Asian conference computer vision, pp 512–520
8. Pillai JK, Patel VM, Chellappa R, Ratha NK (2011) Secure and robust iris recognition using random projections and sparse representations. IEEE Trans Pattern Anal Machine Intell 33:1877–1893
9. Kanatani K (1993) Geometric computation for machine vision. Oxford University Press, Oxford
10. Salo H et al (2015) Spitzer survey of stellar structure in galaxies (S^4G): the pipeline 4: multi-component decomposition strategies and data release. Astrophys J Suppl Ser 219(1)
11. Satriya T, Wibirama S, Ardiyanto I (2016) Robust pupil tracking algorithm based on ellipse fitting. In: 2016 International symposium on electronics and smart devices (ISESD) Bandung, Indonesia pp 253–257
12. Vincze M (2001) Robust tracking of ellipses at frame rate. Pattern Recogn 34:487–498
13. Yahya A, Nordin MJ (2008) A new technique for iris localization in iris recognition systems. J Inf Technol 7:924–929

ChiBa—A Chirrup and Bark Detection System for Urban Environment

Shuddhashil Ganguly, Himadri Mukherjee, Ankita Dhar, Matteo Marciano, and Kaushik Roy

Abstract The World is developing at a tremendous pace which has been catapulted by large-scale technological advancements. Building mega structures has never been easier and modes of commute have also developed thereby shortening travel-time. Such advancements have also brought along newer sources of pollution which are harming our planet at an even faster pace. Sound pollution is one such agent that has a long-term effect on not only humans but the entire biodiversity. Its effect on life is not immediately observed but the damage becomes visible over time. Birds are one of the most affected creatures due to sound pollution. This is one of the major reasons for declining bird population in the Urban areas. It is very important to preserve biodiversity for a sustainable future. Animals have calls that are melodious and rhythmic and these calls tend to change when they are in distress. An automated system can be very useful in this context which can monitor animal sounds and detect changes in their calls. Deployment of such a system in Urban areas is challenging due to the presence of ambient sounds which is extremely diverse. Thus it is essential to initially detect animal calls in the Urban environment prior to monitoring them. ChiBa is a system proposed to address this problem. Experiments were initially performed with the detection of birds and dogs (the most common and loudest creatures in cities) calls in the Urban environment. Tests were performed with over $7K$ clips comprising

S. Ganguly (✉) · A. Dhar
Department of Computer Science and Engineering, Sister Nivedita University, Newtown, West Bengal, India
e-mail: shuddhashil.ganguly@gmail.com

A. Dhar
e-mail: ankita.ankie@gmail.com

H. Mukherjee · K. Roy
TISA Lab, Department of Computer Science, West Bengal State University, Berunanpukuria, West Bengal, India

M. Marciano
Gazelien Records Lab, Department of Arts and Humanities, Music Program, New York University Abu Dhabi, Abu Dhabi, United Arab Emirates
e-mail: matteo.marciano@nyu.edu

© The Author(s), under exclusive license to Springer Nature Singapore Pte Ltd. 2024
D. Giri et al. (eds.), *Proceedings of the Tenth International Conference on Mathematics and Computing*, Lecture Notes in Networks and Systems 963,
https://doi.org/10.1007/978-981-97-2069-9_16

221

of the animal calls as well as Urban ambient sounds. The audios were modeled using a deep learning-based approach wherein the highest accuracy of 99.91% was obtained.

Keywords Sound pollution · Urban ambiance · Bird and dog calls · Deep learning

1 Introduction

The large-scale development in every aspect across the Globe has made our life more comfortable and easier in multitudinous spheres ranging from construction to travel and daily chores. Such advances have also brought along new sources of pollution and our planet is getting harmed at a tremendous rate. Amidst different types of pollution, sound/ noise pollution is an extremely serious issue. It does not demonstrate any immediate/direct effect but acts like a silent killer. Not only humans but the entire biodiversity especially birds are harmed by sound pollution. The large amounts of noise in the Urban areas have had a devastating effect on the animals. Animals have unique calls that are melodious and rhythmic. Their calls change when in distress thus giving a cue toward the presence of pollutants. Monitoring their calls (ecological audio monitoring) can help in this regard. It plays a crucial role in understanding and preserving the delicate balance of our environment. The amount of data that needs to be monitored is very large and thus demands automated systems. One of the major challenges for automated ecological monitoring in Urban ambiance is the identification of the ecological sounds amidst the multifarious sound sources of busy city life. It is essential to distinguish the ecological sounds at the outset prior to monitoring them.

There have been several attempts of distinguishing animal sounds. Oswald et al. [18] have discussed different aspects involved in the distinction of animal sounds. They have also suggested ways of naming animal sounds based on disparate criteria. Nolasco et al. [17] used a few-shot learning-based approach for animal sound classification. The system was designed to work with as low as 5 instances. Experiments were performed on multiple datasets whose details are presented in [17]. Nanni et al. [14] proposed a method to classify animal sounds by combining the dissimilarity of spaces produced by a Siamese neural network. Experiments were performed on a dataset of 2 animals: birds and cats having 2762 and 2962 samples respectively which were parameterized using spectrogram. The spectrograms extracted were fed to the SNN, and the similarity spaces were passed on to an SVM for classification. Ying et al. [11] presented a technique to distinguish animal sounds using doubled-featured spectrogram. They used Local binary pattern variance (LBPV) features, extracted from spectrograms to train a Random forest-based classifier and achieved over 80% accuracy.

Scientists have attempted to distinguish Urban sounds using disparate techniques. Nogueira et al. [16] have discussed various advances in the area of Urban sound classification. They have resented different techniques and available datasets as well as established results on different public datasets. Lezhenin et al. [10] attempted to distinguish Urban sounds using an LSTM-based technique. Experiments were performed on the UrbanSound8K dataset consisting of 8732 clips from 10 classes of length $\leq 4\text{s}^1$. The audios were parameterized using magnitude mel-spectrograms. The system was evaluated using a 5-fold cross-validation technique wherein the highest average precision of 0.85 was reported. Bubashait and Hewahi [3] experimented with CNN, LSTM, and DNN for distinguishing Urban sounds. Experiments were performed on the UrbanSound8K dataset which was parameterized using spectrograms. The spectrograms were fed into different networks and LSTM produced a higher accuracy of 90.5% as compared to the 87.15% accuracy of CNN. Harshavardhan and Mahesh [8] presented an ANN-based approach coupled with MFCC features for urban sound classification. Experiments were performed on the UrbanSound8K dataset with a train-test split of 8 : 2 and the highest accuracy of 87% was reported. Massoudi et al. [13] used a CNN-based approach coupled with Mel-spectrogram for distinguishing Urban sounds from the previous dataset and reported an accuracy of 91%. Das et al. [5] used an LSTM-based approach to distinguish Urban sounds from the UrbanSound8K dataset and reported an accuracy of 98.81% using data augmentation, which was higher than a CNN-based approach. Shu et al. [21] used a CNN-based approach for Urban sound classification using the UrbanSound8K dataset. Their technique involved data augmentation, which was done using a multiple-width frequency-delta-based technique wherein the width was varied from 3–9 with a step of 2. They concluded that data augmentation did not play a major role in the system's performance. However, the segmentation window length greatly influenced the performance. Luitel et al. [12] proposed a method for classifying different urban sounds consisting of bus engines, car horns, and whistles which were parameterized using MFCC-based features. The features were fed to a two-stage classification system comprising of ANN, Naive-Bayes classifier, Decision tree, and Random forest. The highest accuracy of 96.41% was reported in the experiments. Salamon et al. [20] proposed a method of Urban sound classification using unsupervised learning. Experiments were performed on the UrbanSound8K dataset. They used spherical k-means to learn the features and then fed the features into a Random Forest-based classifier with 500 trees. A performance improvement of 5% was reported over an MFCC-based baseline system.

Distinguishing animal calls and monitoring them is challenging due to the overlapping frequency envelope of Urban sounds and animal calls. ChiBa is a system aimed toward the distinction of animal calls (calls of birds (**Chirrups**) and dogs (**Barks**)) from different sounds in the Urban environment. The details of the experiment are presented in the subsequent paragraphs.

---

1 https://urbansounddataset.weebly.com/urbansound8k.html.

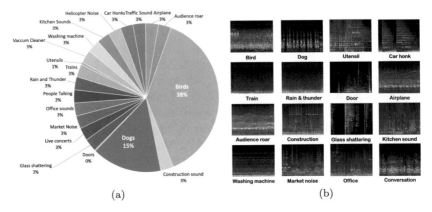

Fig. 1 **a** Distribution of different sound sources in the dataset. **b** Mel-spectrograms for different types of audio in the dataset

2   Dataset Description

The dataset primarily comprises various environmental sounds collected from the Internet, which have been categorized into two groups. The first group comprises sounds containing birds and dogs vocalizations, capturing the natural and harmonious melodies of these creatures (most common animal sounds in the Urban environment). The second group encompasses a wide range of other environmental sounds typically encountered in urban ambiance, providing a comprehensive representation of the various auditory experiences found in bustling cityscapes.

Initially, distinct sounds of dogs and birds, capturing their unique calls, chirps, barks, and songs were collected from the Internet. This formed the first class comprising of 3794 instances, each of length 5 s. The second group of Urban ambient sounds was composed of disparate indoor and outdoor sources. The indoor sources consisted of sounds from washing machines, the whir of vacuum cleaners, the clatter of utensils, doors opening and closing, the hum of kitchen activity, glass shattering with a resonating impact, the ambient sounds of offices, and people engaged in conversations. The external source encompasses a myriad of sounds such as airplanes soaring through the skies, audiences cheering in excitement, car honking, construction activities resonating with urban development, helicopters flying in the sky, the lively atmosphere of concerts, the clamor of bustling markets, the rhythmic patter of rain and the rumble of thunder, the continuous flow of traffic and the roaring of trains. This group totaled to 3246, 5 second long clips. The distribution of different sound sources in the dataset is presented in Fig. 1a. mel-spectrogram representation of some of the audio types is presented in Fig. 1b.

3 Proposed Method

3.1 Feature Extraction

Each of the 5-second clips was used to generate mel-spectrogram [22] which is a variant of the standard spectrogram [25]. A spectrogram is a visual representation of the frequencies present in a signal over time. It is obtained by dividing the audio signal into short time frames and applying the Fourier transform [4] to each frame. The resulting spectrogram represents the magnitude of different frequency components as a function of time. Mathematically, the spectrogram $S(t, f)$ can be expressed as

$$S(t, f) = |STFT(x(t), f)|^2, \tag{1}$$

where $x(t)$ represents the audio signal, f denotes the frequency, and $STFT$ denotes the Short-Time Fourier Transform. The magnitude squared of the $STFT$ is computed to obtain the spectrogram.

The mel spectrogram is a variant of this spectrogram that emphasizes perceptually relevant frequency components. It uses the mel-scale, which is a perceptual scale of pitches that corresponds more closely to human perception of sound than the linear frequency scale. It is calculated by applying a mel filter bank to the spectrogram. Mathematically, the mel-spectrogram $M(t, m)$ can be expressed as

$$M(t, m) = \sum [S(t, f) * H(m, f)], \tag{2}$$

where $H(m, f)$ represents the mel filterbank, and the summation is performed over all frequency bins f. The mel filterbank coefficients are used to weigh the spectrogram values at different frequencies, emphasizing the mel-scale frequencies.

While creating the mel-spectrograms, various windowing functions: Hann [1], Hamming [2], Bartlett [19], Blackman [19], and Bohman [6], were employed to enhance the analysis of the audio signals. This helped to create a comprehensive visual representation of the audio signals for the two major urban life animal sounds birds and dogs, and various other environmental sounds.

3.2 Deep Learning-Based Classification

Deep learning [9] is a branch of machine learning that uses artificial neural networks. These networks consist of multiple layers of interconnected nodes, enabling them to grasp intricate data representations and relationships by identifying patterns and features within the data.

The Convolutional Neural Network [7] is a type of deep learning algorithm that is primarily used for analyzing visual data. CNNs are widely used in computer vision tasks. The architecture of a CNN is designed to automatically extract relevant features from the input data by utilizing convolutional layers, pooling layers, and fully connected layers which are detailed as follows:

- **Convolutional layer**: This layer performs a series of convolutions, where small filters or kernels are applied to the input data to extract local patterns or features. These filters slide over the input, computing dot products at each position and generating feature maps.
- **Pooling layer**: The pooling layer reduces the spatial dimensions of the feature maps by downsampling them. The pooling operation used here is max pooling, which selects the maximum value within a specific region and discards the rest, thus retaining the most prominent features.
- **Fully Connected/Dense layer**: These layers are typically present toward the end of the CNN architecture. They connect every neuron in one layer to every neuron in the next layer, performing classification or regression tasks based on the learned features.

The mel-spectrograms were initially resized to a dimension of $100 * 100$ and fed to the network wherein the first convolution layer was composed of 64, $2 * 2$ filters. The output from this layer was max-pooled and fed to another convolution layer consisting of 32, $3 * 3$ filters. The output from this layer was again max-pooled and finally passed on to the last convolution layer consisting of 16, $4 * 4$ filters. In all the pooling layers, a filter size of $3 * 3$ was used. The output from the final convolution layer was max-pooled and subjected to 50% dropout which was then passed on to a 50-dimensional dense layer followed by a 2-dimensional fully connected layer. All the convolution layers and the initial dense layer were equipped with ReLU activation while the output layer had Softmax activation. The model was trained based on categorical cross-entropy loss function [23] and Adam-based optimization [24]. The number of generated parameters in each of the layers is presented in Table 1.

Table 1 Number of parameters in different layers for the current network

Layer	#Parameters
Convolution-1	832
Convolution-2	18464
Convolution-3	8208
Dense-1	115250
Output	102
Total	142856

Table 2 Accuracy for different windowing functions

Window	**Bartlett**	Blackman	Bohman	Hamming	Hann
Accuracy (%)	**99.81**	99.76	99.58	99.76	99.62

Table 3 Interclass confusions for Bartlett window using *image size* 100 * 100 and *number of training epochs* 500

	Animal sounds	Ambient sounds
Animal sounds	1160	3
Ambient sounds	1	963

4 Results and Analysis

Initially, each of the feature sets (Mel-spectrograms using different windowing functions) was fed to the CNN with an initial image size of $100 * 100$ along with 500 training epochs. The dataset was split into train and test sets in the ratio 7 : 3. The accuracies for the different windowing functions are presented in Table 2.

It is noted that the best result was obtained by using the Bartlett window function, which achieved the highest accuracy. The inter-class confusion for this scenario is presented in Table 3.

It can be observed from Table 3 that 99.74% instances from the animal sounds were correctly classified while 99.90% of the ambient sounds were correctly predicted. Next, considering the Bartlett window function, as it yielded the highest accuracy, different training epochs were used and accuracies were achieved, keeping the image dimension the same as before ($100 * 100$). Different epoch values of 250, 500, 750, and 1000 were used whose results are presented in Table 4.

The best result was obtained for epochs 500 and 750 and no improvement was observed as compared to the initial scenario. Further experimentations were undertaken with the lower number of iterations (500) and the image dimensions were varied as $50 * 50$, $100 * 100$ (initial), $150 * 150$, and $200 * 200$. The obtained accuracies are presented in Table 5.

Table 4 Accuracy for different epochs using 100 dimensional images and Bartlett window

Epochs	250	**500**	**750**	1000
Accuracy (%)	99.76	**99.81**	**99.81**	99.76

Table 5 Accuracy for different image dimensions for 500 epochs and Bartlett window

Image size	50	100	**150**	200
Accuracy (%)	99.44	99.81	**99.91**	99.67

Table 6 Confusion matrix for 150-dimensional images with 500 training epochs and Bartlett window

	Animal sounds	Ambient sounds
Animal sounds	1162	1
Ambient sounds	1	963

Table 7 Performance of different classifiers on Mel-spectrogram computed using Bartlett window

Classifier	RF	SVM	MLP	NB	CNN
Accuracy (%)	98.35	97.12	94.24	86.78	99.91

It is noted that the highest accuracy of 99.91% was obtained for $150 * 150$ dimensional image. In this case, the Bartlett window function was used along with 500 training epochs. The inter-class confusion and accuracy are presented for it in Table 6.

It is noted that the recognition accuracy for the ambient sounds was the same as that of the initial setup. However, in the case of animal sounds, the recognition accuracy improved to 99.91% over the initial 99.74%. The performance of traditional classifiers- Random Forest (RF) [15], Support Vector Machine (SVM) [15], Multilayer Perceptron(MLP) [15], and Naive-Bayes (NB) [15] was also tested for the mel-spectrograms whose results are listed in Table 7. It is noted that CNN was the best performer followed by Random Forest and SVM whose performance was 1.6% and 2.8% below CNN.

5 Conclusion

ChiBa is a system, presented toward the distinction of animal calls from ambient sounds in the Urban environment. Experiments were performed with birds and dog calls (chirrup and bark respectively) along with an array of commonly occurring sounds in the Urban environment. The system was able to distinguish the 2 classes with a mean precision of 99.91%. In the future, the system will be tested on a dataset of larger size wherein greater inter-class variety will be introduced. The dataset will also be subjected to additional noise to test its robustness. Experiments will be performed with handcrafted features for more efficient modeling of the audios to build a handcrafted feature-based system that is at par with deep learning-based systems. The system will also be extended to monitor changes in the pattern of calls of the animals and attempts will be made to perform intra-class distinction. Finally, the deep learning-based architecture will be refined to produce a lighter architecture to facilitate deployment in resource-constrained scenarios which can also aid in real-time processing.

References

1. Barros J, Diego RI (2005) On the use of the Hanning window for harmonic analysis in the standard framework. IEEE Trans Power Deliv 21(1):538–539
2. Bojkovic ZS, Bakmaz BM, Bakmaz MR (2017) Hamming window to the digital world. Proc IEEE 105(6):1185–1190
3. Bubashait M, Hewahi N (2021) Urban sound classification using DNN, CNN & LSTM a comparative approach. In: 2021 International conference on innovation and intelligence for informatics, computing, and technologies (3ICT). IEEE, pp 46–50
4. Cochran WT, Cooley JW, Favin DL, Helms HD, Kaenel RA, Lang WW, Maling GC, Nelson DE, Rader CM, Welch PD (1967) What is the fast fourier transform? Proc IEEE 55(10):1664–1674
5. Das JK, Ghosh A, Pal AK, Dutta S, Chakrabarty A (2020) Urban sound classification using convolutional neural network and long short term memory based on multiple features. In: 2020 Fourth international conference on intelligent computing in data sciences (ICDS). IEEE, pp 1–9
6. Dowd AV, Thanos MD et al (2000) Vector motion processing using spectral windows. IEEE Control Syst Mag 20(5):8–19
7. Gu J, Wang Z, Kuen J, Ma L, Shahroudy A, Shuai B, Liu T, Wang X, Wang G, Cai J et al (2018) Recent advances in convolutional neural networks. Pattern Recogn 77:354–377
8. Harshavardhan K et al (2022) Urban sound classification using ann. In: 2022 International interdisciplinary humanitarian conference for sustainability (IIHC). IEEE, pp 1475–1480
9. LeCun Y, Bengio Y, Hinton G (2015) Deep learning. Nature 521(7553):436–444
10. Lezhenin I, Bogach N, Pyshkin E (2019) Urban sound classification using long short-term memory neural network. In: Federated conference on computer science and information systems (FedCSIS). IEEE, pp 57–60
11. Li Y, Huang H, Wu Z (2019) Animal sound recognition based on double feature of spectrogram. Chin J Electron 28(4):667–673
12. Luitel B, Murthy YS, Koolagudi SG (2016) Sound event detection in urban soundscape using two-level classification. In: IEEE distributed computing, VLSI, electrical circuits and robotics (DISCOVER). IEEE, pp 259–263
13. Massoudi M, Verma S, Jain R (2021) Urban sound classification using CNN. In: 6th International conference on inventive computation technologies (ICICT). IEEE, pp 583–589
14. Nanni L, Brahnam S, Lumini A, Maguolo G (2020) Animal sound classification using dissimilarity spaces. Appl Sci 10(23):8578
15. Nayeem MJ, Rana S, Alam F, Rahman MA (2021) Prediction of hepatitis disease using k-nearest neighbors, Naive Bayes, support vector machine, multi-layer perceptron and random forest. In: International conference on information and communication technology for sustainable development (ICICT4SD). IEEE, pp 280–284
16. Nogueira AFR, Oliveira HS, Machado JJ, Tavares JMR (2022) Sound classification and processing of urban environments: a systematic literature review. Sensors 22(22):8608
17. Nolasco I, Singh S, Morfi V, Lostanlen V, Strandburg-Peshkin A, Vidaña-Vila E, Gill L, Pamuła H, Whitehead H, Kiskin I et al (2023) Learning to detect an animal sound from five examples. arXiv:2305.13210
18. Oswald JN, Erbe C, Gannon WL, Madhusudhana S, Thomas JA (2022) Detection and classification methods for animal sounds. In: Exploring animal behavior through sound, vol 1, pp 269–317
19. Putranto P, Desvasari W, Daud P, Wijayanto YN, Mahmudin D, Kurniadi DP, Rahman AN, Hardiati S, Setiawan A, Darwis F et al (2019) Performance comparison of Blackman, Bartlett, Hanning, and Kaiser window for radar digital signal processing. In: 4th International conference on information technology, information systems and electrical engineering (ICITISEE). IEEE, pp 391–394

20. Salamon J, Bello JP (2015) Unsupervised feature learning for urban sound classification. In: IEEE international conference on acoustics, speech and signal processing (ICASSP). IEEE, pp 171–175
21. Shu H, Song Y, Zhou H (2018) Time-frequency performance study on urban sound classification with convolutional neural network. In: TENCON 2018–2018 IEEE region 10 conference. IEEE, pp 1713–1717
22. Zhang T, Feng G, Liang J, An T (2021) Acoustic scene classification based on Mel spectrogram decomposition and model merging. Appl Acoust 182:108258
23. Zhang Z, Sabuncu M (2018) Generalized cross entropy loss for training deep neural networks with noisy labels. In: Advances in neural information processing systems, vol 31
24. Zhang Z (2018) Improved adam optimizer for deep neural networks. In: IEEE/ACM 26th international symposium on quality of service (IWQoS). IEEE, pp 1–2
25. Zue V, Cole R (1979) Experiments on spectrogram reading. In: ICASSP'79 IEEE international conference on acoustics, speech, and signal processing, vol 4. IEEE, pp 116–119

Guarding the Beats by Defending Resource Depletion Attacks on Implantable Cardioverter Defibrillators

Anisha Mitra and Dipanwita Roy Chowdhury

Abstract Implantable Medical Devices (IMDs) have revolutionized the treatment of critical diseases. However, the increasing reliance on these life-saving devices' wireless functionality has made them vulnerable to cyber-attacks. Implantable Cardioverter Defibrillator (ICD) has emerged as a leading IMD owing to the worldwide surge in cardiac diseases. Given the resource-constrained ICD environment, there's a pressing need to develop tailored security measures for protection, moving beyond traditional approaches. In this paper, we present resource depletion attack scenarios in an ICD environment where attackers can exploit the ICD's wireless connectivity function. We propose some comprehensive approaches to mitigate such attacks, offering a significant step forward in safeguarding the well-being of patients. This research contributes to the ongoing efforts to secure the Internet of Medical Things (IoMT) ecosystem and underscores the importance of cyber-security in modern healthcare.

Keywords Resource depletion attack · Wireless communication · Resource constraint device · Security of ICD

1 Introduction

The worldwide surge in cyber-physical system security encompasses the vital aspect of safeguarding medical implants and ensuring patient safety and data integrity. Implantable medical devices (IMDs) are electronic medical gadgets that are surgically implanted inside the body to diagnose, monitor, treat, or support various medical conditions. These devices can vary in complexity and function, still, they generally serve purposes like heart rhythm management, drug delivery, neurostimu-

A. Mitra (✉) · D. R. Chowdhury
Department of Computer Science and Engineering, Indian Institute of Technology Kharagpur, Kharagpur, West Bengal, India
e-mail: mitraanisha.15@gmail.com

D. R. Chowdhury
e-mail: drc@cse.iitkgp.ac.in

lation, visual or hearing aid, etc. Wireless connectivity made it possible for doctors and patients to remotely monitor IMDs by connecting two devices or by using the internet. Medical Implant Communication Service (MICS) is a low-power, short-range (2m), high data-rate communication band that operates between 401 and 406 MHz [10]. Wireless communication characteristics can be misused to undermine the security of IMDs, despite their many advantages. A vulnerable IMD environment in the presence of multiple external interfaces demands relevant security measures to protect precious human lives. Several cyber-security threats can hamper legitimate IMD operations, misuse patient-specific private treatment logs, and deplete battery power unnecessarily. However, there are certain limitations to the functioning of IMDs in constrained resources with very little memory, small controller, and battery-powered. The resource-constrained IMD environment and their unfamiliar access requirements during device or patient emergencies make the adoption of traditional security approaches impractical in this domain.

Implantable Cardioverter Defibrillators (ICDs) are one of the most active IMDs due to the global increase in cardiac diseases. Along with an external programmer control, ICDs support remote monitoring from the patient's place using a home monitor device and data exchange with a cloud-based hospital network. They interface with cloud-based systems using Internet protocol and home monitoring devices, and they communicate with outside programmers and home monitors using Radio Frequency (RF) communication signals sent in the MICS band [15].

Resource depletion is identified as one of the most dangerous attacks on the ICD environment as it reduces an ICD's lifetime from a few years to a few months. ICDs Lithium batteries are very much vulnerable to resource depletion attacks and have no option of recharging unlike general medical sensors' batteries [9]. Thus, thwarting resource depletion without a battery replacement is a task of significant importance for ICDs.

In this paper, we focused on resource depletion attacks on the ICD environment and depicted some attack scenarios based on existing research documents and ICD manuals. The rest of the paper is organized as follows. In Sect. 2, we discuss the ICDs working principle as well as the communication environment and related works on IMD security. Section 3 depicts the proposed attack models on the ICD environment for resource depletion attacks. In Sect. 4, we present some relevant countermeasures for identified attacks and conclude in Sect. 5.

The novelty of this paper lies in the identification of IMD's environment-specific attack scenarios and proposals for relevant countermeasures.

Our Contribution

- Considering the real-time IMD environment, definite resource depletion attack scenarios are identified, and their algorithmic representations are included very precisely.
- Countermeasures to thwart detected attack scenarios are proposed specifically in relevance to extreme resource constraint IMD environment.

2 Preliminaries

This section summarizes details about market-available ICDs and discusses the existing works on resource depletion attacks in the ICD environment. Current research progress on ICD security as well as market-available ICD manuals of leading ICD manufacturers play the role of primary resources for this work.

2.1 *Implantable Cardioverter Defibrillators (ICD)*

An Implantable Cardioverter Defibrillator (ICD) is an electronic device that constantly monitors heart rhythm. Generally, when an ICD senses a rapid heartbeat, it can administer an electrical shock to restore a normal heart rhythm.

An ICD consists of a pulse generator and electrodes or leads that pass through the veins and connect the heart with the ICD. ICDs work similarly to pacemakers, but instead of treating slow heart rhythms, they use shock therapy or fast-paced impulses to correct abnormal heart rhythms. There exists a subcutaneous version of ICD (S-ICD). It has an electrode in the tissue of the left side of the breastbone and can only give high-energy shocks. S-ICD does not have a pacemaker function to speed up or slow down heart rhythms.

Post-implant, medical personnel utilize an external programmer to adjust ICDs therapy configuration, retrieve patient health records, and modify device parameters for accurate treatment delivery. This human control if maliciously used can lead to serious health hazards due to inaction (failure to deliver necessary treatment) or unnecessary shock delivery [7].

2.2 *ICD Communication*

Current ICD models support both short-range and long-range communication for telemetry between the programmer and ICD as well as between the home monitor and ICD. The primary goal of this communication is to update therapy parameters and store patients' regular health data. For short-range communication, the pacing device uses inductive RF communication(0–300 kHz) to initiate telemetry sessions within a short distance(0–10cm) using a magnetic field. A magnetic switch in the implanted ICD activates wireless communication when the magnetic head of the programmer is in proximity. Thus, it is called *wand-telemetry* [7, 11, 15]. For long-range communication without magnet usage, ICDs support RF link telemetry using a radiating RF field which does not require any device proximity conditions. This provides long-range (0–200m) telemetry support transmitted at MICS frequency range (402–405 MHz). This is known as *wand-less telemetry* [15]. However, this telemetry service can get disrupted due to interference with other RF-enabled nearby devices that operate at frequencies near that of pulse generators [3].

Although the wireless link used for this telemetry facility enables remote ICD monitoring, which improves the delivery of care, it may also be used by adversaries to undertake resource depletion assaults, as shown in the following sections.

2.3 Existing Literature Review

Decades-long research progress in the IMD security domain presented some good and detailed works. Pirretti et al. in [17] depicted sleep deprivation and barrage attacks on sensor network nodes and proposed three mitigation schemes, highlighting their applicability for sensor-based IMD functionalities but not directly. Research works in [5, 12, 16], discovered the significance of resource depletion attacks in wireless mobile devices and battery-powered mobile computers respectively. Martin et al. in [12] proposed three different types of Sleep deprivation attacks on battery-powered mobile computers along with a power-secure architecture to thwart such attacks. The attack scenarios under consideration bear a resemblance to the challenges encountered in IMD's resource-constrained environment, but secure architecture does not fit there.

In one of the existing state-of-the-art research works in the resource depletion attack domain for IMDs [9], the authors proposed a machine learning model (SVM) based countermeasure. This work involves the patient's smartphone as a line of defense before the underlying authentication scheme works. This much-appreciated work utilizes patient IMD usage pattern data. Gaining access to patient-specific private data poses a significant challenge, specifically in developing nations like India, China, etc., where expensive IMD usage is not common. The major drawback of the work lies in the usage of a time-consuming machine learning model and the inclusion of another attack-prone smartphone device in the security scheme. In comparison to that, our work presents more elaborate and IMD environment-specific resource depletion scenarios as well as efficient and relevant countermeasures without involving any other device.

Kevin Fu et al., in [7] used general-purpose software radios to stimulate unidentified attacks on the IMD environment, revealing the fact that attackers can reverse engineer proprietary protocols. They captured and processed RF traces of IMD communication and utilized those RF traces later to launch replay attacks. They addressed security risks using human perceptible and zero-power mitigation techniques. However, they only mentioned possible power depletion attacks and did not elaborately depict resource depletion attack scenarios in their work.

Preneel et al. in [11] performed reverse engineering from a weak attacker's approach using off-the-shelf commercial equipment. They have proposed a key agreement protocol using bilinearity and non-degeneracy involving huge computation which seems to be impractical in resource constraint IMD environment.

In [15], researchers carried out a risk assessment on specifically cardiac implants with actor-based, scenario-based, and combined analysis but no countermeasure was proposed to thwart identified attacks. Ngamboe et al. in [13] exposed the vulnerabilities of pacemakers to radio-based attacks and the feasibility of the attacks under real conditions. In a recent work [14], researchers proposed one of the most dangerous, data manipulation attacks on an ICD environment. They depicted the attacker's approach to modifying legitimate treatment configuration and medical data using a malfunctioning programmer device.

We also explored research works where other specific attack scenarios were discovered and countermeasures were proposed. Some of the works [4, 6, 18, 19] proposed external devices or token-based security measures but failed to secure them from wireless connection-related threats. Some Biometric feature-based security solutions were identified [1, 2, 8] which were most promising for their lightweight approach but due to biometric changes over time, these security models do not provide complete protection.

In our work, we focus on resource depletion attack scenarios in a battery-powered ICD environment.

3 Resource Depletion Attack Models

In this section, we propose four identified resource depletion attack models in an ICD environment where the ICD telemetry communication facility can be exploited by an adversary to deplete ICDs' power source rapidly. We propose certain attack scenarios that can arise in an ICD environment due to the use of malfunctioning off-the-shelf equipment. Four such attack scenarios are identified by observing the functioning environment of ICDs which are discussed below considering certain conditions for each.

3.1 Attacks on Low Power Modes of ICD

For sensor-based devices, power management plays an important role as such devices work based on limited battery resources, and most of those are equipped with specific sleep modes depending on the activation schedule. As discussed earlier, ICD's resource constraint environment does not allow recharging of existing batteries due to the direct involvement of the human body (implant). However, the ICD environment may be subjected to similar kinds of attacks as observed in sensor networks where attackers try to keep the victim ICD out of its power-conserving sleep mode [17].

Research findings suggest that Modern ICDs operate in three specific modes: Interrogation, Reprogramming, and Test for active telemetry sessions. Additionally, ICDs support Sleep, and Standby modes for low-power device functioning. The ICD is in 'sleep' mode while there isn't an active telemetry session going on, and in this mode, it periodically checks for wireless inbound access requests from the programmer or home monitor. When an active telemetry session ends, ICD slips to 'standby' mode and stays there for five minutes, then goes to sleep mode. During standby mode, the long-range communication channel can be utilized by any programmer to again start an active session by sending a specific message [11]. These two low-power modes of ICD can be exploited by an adversary to launch two different resource depletion scenarios as discussed below:

Barrage Attack: In sensor networks, the victim sensor node is bombarded with genuine access requests and does not let it enter sleep mode by engaging in resource-exhaustive tasks [17]. ICD environment can be a victim of similar kinds of attack scenarios. To access an ICD, an external programmer must perform an authentication procedure and then only telemetry service is enabled. But, an adversary can use an off-the-shelf programmer to send such an authentication request which in turn requires excessive computation from the ICD's side and consumes battery power. Continuous authentication requests will not allow ICD to enter low-power 'sleep' mode as depicted in Algorithm 1. If an attacker sends an authentication request with legitimate credentials and gets access after verification, it does nothing and keeps on sending access requests in a loop. In case, unauthorized access requests get rejected, the attacker lets ICD change its state to 'standby' mode and wait for 5 min. Then again sends an access request, so that ICD cannot enter 'sleep' mode after coming out from 'standby' mode. This results in unnecessary resource depletion if continued for a long duration.

Algorithm 1 Barrage Attack

$i \leftarrow 1; n \leftarrow N;$ ▷ Consider 'N' to be a large value
while TRUE **do**
L1:
    if 'ICD' is in 'SM' **then**   ▷ Sleep Mode: 'SM'
        Attacker $\xrightarrow[\text{with 'M'}]{\text{Access Request}}$ 'ICD'   ▷ 'M' defines Attacker's identification
        for $i = 1, \ldots, n$ **do**   ▷ Checks 'M' for Authentication

            if Authorised **then**
                Release Access Grant
                Attacker $\xrightarrow[\text{with 'M'}]{\text{Access Request}}$ 'ICD'

            else
                delay(5)   ▷ Wait for '5 minutes' for 'standby' mode to end
                goto L1
            end if
        end for
    end if
end while

Sleep Deprivation Attack: Unlike a barrage attack, in a sleep deprivation attack, the attacker does not let ICD leave standby mode. Here as depicted in Algorithm 2, the adversary waits for an active session to end and lets the ICD enter 'standby' mode. Then he/she sends a specific message to activate the ICD but does nothing and the ICD again goes to 'standby' mode. Just after four minutes, before the five-minute 'standby' mode timer expires and the ICD enters 'sleep' mode, the again attacker tries to activate the ICD. Thus, this attack does not let ICD come out of 'standby' mode and enter 'sleep' mode which in turn results in unnecessary battery depletion.

Algorithm 2 Sleep Deprivation Attack

$i \leftarrow 1; n \leftarrow N;$ ▷ Consider 'N' to be a large value
for $i = 1, \ldots, n$ **do**
L2:
    if 'ICD' is in 'SB' **then**   ▷ Standby Mode: 'SB'
        Attacker $\xrightarrow[\text{message}]{\text{A Specific}}$ 'ICD'
        delay(4)   ▷ Wait for '4 minutes' to keep the ICD in 'Standby' mode
        goto L2
    end if
end for

3.2 Prolonged Telemetry Communication

In this attack, ICD's magnet-based short-range communication can be exploited to reduce battery life. As discussed earlier, the wand-telemetry session starts in the ICD environment by placing the magnetic head near the ICD's pulse generator which in turn enables the long-range RF wave-based wireless telemetry. As shown in the attack Algorithm 3, the attacker initiates short-range telemetry by using its own off-the-shelf programmer and magnetic head and keeps it placed near the ICD. After exchanging necessary messages with ICD, RF wireless telemetry communication is enabled. However, the attacker does not exchange any other telemetry message. The wireless telemetry session terminates if the ICD loses communication with the programmer for a prolonged period of one hour. The attacker keeps a strong magnet near the device to keep the device in telemetry interrogation mode continuously. This also leads to resource depletion.

Algorithm 3 Prolonged Telemetry Communication

$i \leftarrow 1; n \leftarrow N;$ ▷ Consider 'N' to be a large value

for $i = 1, \ldots, n$ **do**

    if 'Wand-telemetry' is established **then**

        L3: Replay 'Wand-less telemetry' initiation Command

    if 'Wand-less Telemetry' is enabled **then**

            wait(60)   ▷ Wait for 'one hour'

            goto L3

    end if

  end if

end for

3.3 Electromagnetic Interference on the Wireless Channel (Jamming)

This attack requires the adversary to use special equipment to capture radio frequency signals transmitting between the programmer and ICD. The commercial radio software must be used to identify the ICD's telemetry reception frequency and an antenna should be tuned in that frequency. Considering the availability of the mentioned equipment, the below-mentioned steps may be followed by an attacker to raise the signal power of the device for wireless data transmission which results in resource depletion.

Algorithm 4 Jamming Attack

Step 1: Attacker checks if any active session is going on between IMD and programmer.

Step 2: If no ongoing active session exists, go to **Step 1**.

Step 3: If there is an active session, then the attacker sends high power noise signal at ICD's reception frequency.

Step 4: Disrupts ICD's ongoing session

Step 5: ICD raises signal power to receive legitimate telemetry data.

3.4 Device Functional Parameter Modification

The ICD batteries provide the device power, enabling it to monitor heartbeat and deliver electrical shocks or pulses to treat the heart. Now, most market-available ICDs follow a functional model to provide correct treatment delivery. This func-

Table 1 Malicious device parameter modification

SI. no.	Parameters	Identified modification
1.	Atrium pulse amplitude (P.A.)	P.A. = 7.5
	Left Ventricular pulse amplitude (L.V.A.)	L.V.A. = 7.5
	Right Ventricular pulse amplitude (R.V.A.)	R.V.A. = 7.5
2.	Atrium pulse width (P.W.)	P.W. = 1.5
	Left Ventricular pulse width (L.V.W.)	L.V.W. = 0.4
	Right Ventricular pulse width (R.V.W.)	R.V.W. = 0.4
3.	Maximum shock delivery (maxsh)	maxsh $\gg$ 5

tional model is based on certain programmable parameters which control therapy delivery. Some programmable parameters can affect battery longevity. If an adversary changes such parameters to any incorrect values, it might potentially result in unnecessary battery drain. Pulse amplitude and pulse width are two programmable ICD parameters that may control the battery life of the ICD.

Pulse amplitude: At the leading edge of the output pulse, the voltage of the pulse is measured. The square of the amplitude directly relates to the amount of energy given to the heart. The energy provided is therefore quadrupled by doubling amplitude [3]. Thus, ideally, battery life is extended by lowering the amplitude while preserving a safety margin.

Pulse width: The energy delivered to the heart is directly proportional to the pulse width. So, doubling the pulse width doubles the energy delivered. Similar to the pulse amplitude parameter, battery life may be extended by setting a shorter pulse width while preserving the appropriate safety margin [3].

These two parameters can be modified externally by using a programmer device. Attackers can use external radio software to capture RF signals in which the programmer modifies these parameters. Later, these captures can be replayed with attacker-defined values. A successful replay attack will change these two parameters to the attacker's choice of values which may lead to battery depletion.

Similarly, the number of high-energy shocks delivered to the patient on detection of a legitimate heart condition for a single session affects battery performance. If the attacker sets a high value for this parameter, then on detection of the Ventricular Fibrillation condition, multiple unnecessary high-energy shocks will be delivered. Continuous shock delivery will endanger the patient's life as well as reduce battery life immensely.

Table 1 depicts the possible parameter modifications that can deplete ICD resources irrelevantly.

4 Countermeasures

We have discussed in the previous sections about resource depletion attack scenarios. In this section, we are proposing some countermeasures to thwart detected attacks considering resource constraints in the ICD environment.

– In a real-time medical setting, it is very difficult to adopt traditional security measures involving time-taking computations. Response to access requests to the ICD should be prompt and secure. Algorithms 1 and 2 depicts how an attacker can deliberately force an ICD to perform resource-consuming authentication operations continuously. It blocks ICD from entering low-power modes. We suggest a less time-consuming verification method based on low-cost XOR operation by using a buffer and counter. We assume a 4-bit ID value for the programmer's verification which is denoted by 'M' in Algorithms 1 and 2. Consider, that during the authentication phase, ICD uses a 4-bit buffer and a 2-bit counter. The buffer keeps the ID of the user sending an access request in the most recent past and the counter keeps track of the consecutive requests sent by the same user. The buffer and the ID of the current user 'M' are compared to identify if the two consecutive requests come from the same user. The counter is initialized to 0. Figure 1 depicts the working of the proposed mitigation method. If the comparison is true then it indicates the same user 'M' puts the request access and the counter is increased by 1. A threshold of '3' is used for counter and if crossed it indicates an attack situation using the same credentials. Otherwise, the counter is toggled to zero and the buffer saves 'M'. When an attack situation is detected, an alert signal is generated to notify physicians, and access is blocked.
Considering that the underlying IMD environment is protected using relevant crypto primitives for authentication and encryption, our proposed scheme provides a realistic solution to protect against the proposed resource depletion scenario. This approach does not let unauthorized adversaries access private IMD data by block-

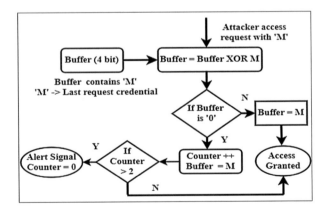

Fig. 1 Proposed countermeasure for attacks on low-power mode

ing multiple continuous authentication requests and thus satisfies authentication and authorization. However, access to IMD services is made available for authenticated personnel by using the mentioned threshold value for the counter. Thus, protecting IMD's access against attackers the confidentiality and integrity of IMD data is satisfied.

- Algorithm 3 depicts a resource depletion attack scenario using continuous wand-telemetry. We have discovered that a wand-less telemetry session without the use of a magnetic wand terminates after a specific duration if communication is lost. After wand-less telemetry is disabled, if a wand-telemetry is detected immediately, a timer can run for a specific duration. After the timer expires, if still wand-telemetry continuous, an alert signal can be used to alert the patient or physician. It can notify them about the possible placement of any powerful magnet other than the programmer head near the ICD which causes unnecessary wand-telemetry communication.
- Jamming-based attacks can be avoided by using possible anti-jamming techniques.
- Unnecessary functional parameter modification can be guarded by securing exchanged telemetry messages between the programmer and ICD. The specific range of values for different parameters allowed for smooth device functioning should be known. To thwart incorrect parameter modification lightweight authenticated encryption algorithms can be used to encrypt ongoing data transmission. So that, improper parameter modification beyond the allowed range can be detected and stopped. The prospective lightweight approach should prioritize minimal memory usage and low computational overhead to align with IMD's resource-limited environment.

5 Conclusion

In this paper, we have discussed definite resource depletion attack scenarios in the ICD environment. We observe that attackers may exploit the wireless communication environment of ICD to disrupt legitimate functionalities. Here, some countermeasures to thwart such attacks are proposed. Existing research works available in the literature, are mostly based on old ICD models. We recognize a genuine need to simulate these attack scenarios based on currently available ICD models in the market. This would ensure the applicability of detected attacks in clinical settings. In the absence of practical attack models, it is necessary to mount defenses against various cyber-attacks in the IMD environment. So, this research domain warrants profound consideration from the relevant research community. Advancements in this research domain hold significance not only for academic purposes but also for the betterment and well-being of human lives, making it a vital area of exploration for young researchers.

References

1. Almukhlifi A, Almutairi S (2023) Efficient palm vein authentication encryption technique in wireless implantable medical devices. Indones J Electr Eng Comput Sci 30:1651
2. Belkhouja T, Du X, Mohamed A, Al-Ali A, Guizani M (2019) Biometric-based authentication scheme for implantable medical devices during emergency situations. Future Gener Comput Syst 98. https://doi.org/10.1016/j.future.2019.02.002
3. Boston Scientific, Marlborough, USA: Boston Scientific reference Guide for ICDs (2017)
4. Denning T, Fu K, Kohno T (2008) Absence makes the heart grow fonder: new directions for implantable medical device security. In: Proceedings of the 3rd conference on hot topics in security. HOTSEC'08, USENIX Association, USA
5. Desnitsky V, Kotenko I, Zakoldaev D (2019) Evaluation of resource exhaustion attacks against wireless mobile devices. Electronics 8(5). https://doi.org/10.3390/electronics8050500, https://www.mdpi.com/2079-9292/8/5/500
6. Gollakota S, Hassanieh H, Ransford B, Katabi D, Fu K (2011) They can hear your heart-beats: Non-invasive security for implantable medical devices. In: Proceedings of the ACM SIGCOMM 2011 conference (SIGCOMM '11). Association for Computing Machinery, New York, NY, USA, pp 2–13. https://doi.org/10.1145/2018436.2018438
7. Halperin D, Heydt-Benjamin TS, Ransford B, Clark SS, Defend B, Morgan W, Fu K, Kohno T, Maisel WH (2008) Pacemakers and implantable cardiac defibrillators: Software radio attacks and zero-power defenses. In: 2008 IEEE symposium on security and privacy (sp 2008), pp 129–142. https://doi.org/10.1109/SP.2008.31
8. Hei X, Du X (2011) Biometric-based two-level secure access control for implantable medical devices during emergencies. In: 2011 Proceedings IEEE INFOCOM, pp 346–350. https://doi.org/10.1109/INFCOM.2011.5935179
9. Hei X, Du X, Wu J, Hu F (2010) Defending resource depletion attacks on implantable medical devices. In: IEEE global telecommunications conference GLOBECOM, pp 1–5
10. Islam MN, Yuce MR (2016) Review of medical implant communication system (mics) band and network. ICT Express 2(4):188–194, special Issue on Emerging Technologies for Medical Diagnostics
11. Marin E, Singelée D, Garcia FD, Chothia T, Willems R, Preneel B (2016) On the (in)security of the latest generation implantable cardiac defibrillators and how to secure them. In: Proceedings of the 32nd annual conference on computer security applications (ACSAC '16). Association for Computing Machinery, New York, NY, USA, pp 226-236. https://doi.org/10.1145/2991079.2991094
12. Martin T, Hsiao M, Ha D, Krishnaswami J (2004) Denial-of-service attacks on battery-powered mobile computers. In: . Proceedings of the second IEEE annual conference on pervasive computing and communications, pp 309–318. https://doi.org/10.1109/PERCOM.2004.1276868
13. Ngamboe M, Fernandez JM, Dyrda K (2021) Radio-based cyber-attacks against pacemakers: assessing their chance of success under real conditions. Cardiol Cardiovasc Med 5:591–598
14. Mitra A, Roy Chowdhury D (2023) Unmasking the dominant threat of data manipulation attack on implantable cardioverter defibrillators. In: Proceedings of the 20th annual international conference on privacy, security and trust (PST). IEEE Xplore

15. Ngamboe M, Berthier P, Nader A, Dyrda K, Fernandez J (2021) Risk assessment of cyber-attacks on telemetry-enabled cardiac implantable electronic devices (CIED). Int J Inf Secur 20:1–25
16. Nguyen VL, Lin PC, Hwang RH (2019) Energy depletion attacks in low power wireless networks. IEEE Access 7:51915–51932. https://doi.org/10.1109/ACCESS.2019.2911424
17. Pirretti M, Zhu S, Narayanan V, McDaniel P, Kandemir M, Brooks R (2006) The sleep deprivation attack in sensor networks: analysis and methods of defense. IJDSN 2:267–287. https://doi.org/10.1080/15501320600642718
18. Xu F, Qin Z, Tan CC, Wang B, Li Q (2011) Imdguard: securing implantable medical devices with the external wearable guardian. In: Proceedings IEEE INFOCOM, pp 1862–1870. https://doi.org/10.1109/INFCOM.2011.5934987
19. Zhang M, Raghunathan A, Jha NK (2013) Medmon: securing medical devices through wireless monitoring and anomaly detection. IEEE Trans Biomed Circuits Syst 7(6):871–881. https://doi.org/10.1109/TBCAS.2013.2245664

Author Index

© The Editor(s) (if applicable) and The Author(s), under exclusive license
to Springer Nature Singapore Pte Ltd. 2024
D. Giri et al. (eds.), *Proceedings of the Tenth International Conference on Mathematics and Computing*, Lecture Notes in Networks and Systems 963,
https://doi.org/10.1007/978-981-97-2069-9

Printed in the United States
by Baker & Taylor Publisher Services